Lei Zhang

Sub-Kelvin scanning tunneling microscopy on magnetic molecules

Experimental Condensed Matter Physics
Band 25

Herausgeber
Physikalisches Institut
Prof. Dr. David Hunger
Prof. Dr. Alexey Ustinov
Prof. Dr. Georg Weiß
Prof. Dr. Wolfgang Wernsdorfer
Prof. Dr. Wulf Wulfhekel

Eine Übersicht aller bisher in dieser Schriftenreihe erschienenen Bände finden Sie am Ende des Buchs.

Sub-Kelvin scanning tunneling microscopy on magnetic molecules

by

Lei Zhang

KIT Scientific Publishing

Dissertation, Karlsruher Institut für Technologie
KIT-Fakultät für Physik

Tag der mündlichen Prüfung: 20. Juli 2012
Referenten: Prof. Dr. W. Wulfhekel, Prof. Dr. Georg Weiß

Impressum

KIT Scientific Publishing

Karlsruher Institut für Technologie (KIT)
KIT Scientific Publishing
Straße am Forum 2
D-76131 Karlsruhe

KIT Scientific Publishing is a registered trademark
of Karlsruhe Institute of Technology.
Reprint using the book cover is not allowed.

www.ksp.kit.edu

Print on Demand 2019 – Gedruckt auf FSC-zertifiziertem Papier

ISSN 2191-9925
ISBN 978-3-86644-950-3
DOI 10.5445/KSP/1000031261

Contents

1 Introduction **1**

2 Theoretical background **3**

2.1 Scanning tunneling microscopy 3

2.1.1 The tunneling effect 3

2.1.2 Bardeen's approach and Tersoff-Hamann model 5

2.1.3 STM operation modes 7

2.1.4 Tunneling spectroscopy 7

2.2 Magnetism 10

2.2.1 Spins in crystal field 10

2.2.2 Antiferromagnetically coupled dimers 15

2.2.3 The Kondo effect 16

3 Experimental setup **21**

3.1 Technical properties 21

3.2 System design 22

3.2.1 Cryostat 22

3.2.2 STM head 35

3.2.3 Samples and STM tips 36

3.2.4 Vacuum system 37

3.2.5 Molecular deposition 37

3.2.6 Electronics 38

3.2.7 Vibration isolation 39

3.2.8 4 T superconducting magnet 39

3.2.9 Temperature control 41

3.3 System test results 41

3.3.1 Z-stability and noise level of tunneling current 42

3.3.2 Lateral thermal drift 45

3.3.3 Flux vortices in Nb(110) 46
3.3.4 Energy resolution 46

4 Bipyrimidine-bridged homobinuclear complexes 49
4.1 (M-(hfacac)$_2$)$_2$-bpym complexes 49
4.2 Magnetic properties 50
4.3 Degassing 52
4.4 Mn_2 on Au(111) 53
4.5 Mn_2 on Cu(100) 54
4.5.1 Deposition and adsorption 54
4.5.2 STS of Mn_2/Cu(100) 55
4.5.3 STS map 58
4.5.4 A brief summary 58
4.6 Ni_2 on Cu(100) 58
4.6.1 Deposition and adsorption 59
4.6.2 STS of Ni_2 on Cu(100) 61
4.6.3 A brief summary 65
4.7 Ni_2 on CuN/Cu(100) 66
4.7.1 Growth of the CuN thin film and deposition of Ni_2 67
4.7.2 STS of Ni_2 on CuN/Cu(100) 68
4.8 Zn_2 on Cu(100) 70
4.9 Conclusions and discussions 71

5 Metal Acetylacetonates 73
5.1 Molecular structures 73
5.2 Experimental setup 74
5.3 Dy(acac)$_3$ on Cu(111) 74
5.3.1 Degasing, deposition and adsorption 74
5.3.2 STS 76
5.4 Cr(acac)$_3$ on Cu(111) 79
5.5 Summary 82

6 Conclusions and outlook 83

Deutsche Zusammenfassung 85

Acknowledgements 99

1 Introduction

In the last decades, as predicted by Moore's and Kryder's law [1, 2], semiconductor processing has been remarkably improved and the areal density of magnetic information storage devices has been increased rapidly. Recently, processors produced by 22 nm technology [3] and hard disks with a storage density of 620 GBit per square inch [4], which corresponds to a bit size of in 30x30 nm^2, have been launched to the market. However these development trends cannot hold forever because of the physical limits. In the foreseeable future devices with a length scale of single atoms or molecules will be needed, which, unfortunately, cannot be accomplished by present-day industrial technologies. Therefore, magnetic molecules, especially single molecular magnets (SMMs), have attracted interest. They could behave as single molecular magnetic bits [5] and furthermore exhibit rich potential applications in molecular spintronics [6], such as quantum computing [7].

For applications like magnetic data storage, the magnetic molecules have to be deposited onto a supporting substrate. To realize a magnetic bit on a surface, an intrinsic remnant magnetization of the SMMs, two stable magnetic ground states as well as an energy barrier, the so-called magnetic anisotropy energy (MAE) separating the two ground states, are required. Furthermore the MAE should be sufficiently large so that the magnetic bits can be thermally stable. However, some higher order effects, such as the Kondo effect, can appear in the system due to the hybridization of the molecular states with the electrons of the substrate [8]. In this effect the spin of SMMs is screened by the conduction electrons of the substrate [9] and the magnetic ground state of the SMMs is lost. Thus, in order to store information in SMMs, the types of the SMMs as well as the substrate need to be chosen carefully, so that the Kondo effect can be avoided.

For the study of magnetic molecules on a substrate, an instrument with high lateral resolution and high energy resolution for electronic states is needed. A usual solution for this requirement is a scanning tunneling microscope (STM) [10, 11, 12] working at cryogenic temperatures [13, 14, 15, 16, 17, 18, 19, 20]. For surface science studies of atomically clean structures, the STM needs to be operated in ultra-high vacuum (UHV)

[12, 21, 22, 23]. By performing scanning tunneling spectroscopy (STS), the local density of states (LDOS) of the investigated objects can be obtained [12]. Furthermore, the energy barrier of excitation processes, such as spin-flip, phonon excitation as well as the MAE can be extracted by performing inelastic tunneling spectroscopy (ITS). Typical commercially available low temperature STMs working in UHV normally use He-bath cryostats and operate at $T \geq 4.2$ K [24, 25, 26]. The energy resolution of ITS at the temperature T is limited to 5.4 $k_B T$ [27] which is only about 1.9 meV at 4.2 K. For many interesting questions such as magnetic excitations of single atoms and molecules [28, 29, 30], this resolution is insufficient and lower temperatures are required which can be achieved by using a Joule-Thomson refrigerator or a dilution refrigerator.

In this work, the magnetic behavior of different magnetic molecules on substrates, especially the exchange interaction between the magnetic ions, the magnetic anisotropy on the surface, the magnetic excitations as well as the Kondo effect, were studied by using scanning tunneling microscopy (STM). The theoretical background of the spin-flip process of magnetic molecules in a crystal field and the Kondo effect as well as an introduction to STM and STS, are given in Chapter 2. An ultra stable and low noise STM operating at 400 mK using ^{3}He (930 mK using ^{4}He) has been developed and used to measure the magnetic molecules. An introduction to this set-up is shown in Chapter 3. Magnetic molecules often contain more than one magnetic center, which couple ferromagnetically or antiferromagnetically via exchange interaction [31]. In the simplest case with antiferromagnetically coupled magnetic centers, the magnetic molecule contain two magnetic ions. In Chapter 4 a series of binuclear metal-organic molecules of the form (M(hexafluoroacetylacetonate)$_2$)$_2$bipyrimidine , where M=Nickel, Manganese and Zinc, are investigated in order to measure the strength of the antiferromagnetic coupling between the two transition metal ions. The metal acetylacetonates complexes are another series of coordination complexes, which have been intensively studied [32, 33, 34, 35, 36, 37, 38, 39, 40], and the dysprosium acetylacetonate (Dy(acac)$_3$) shows a relaxation effect corresponding to a spin-lattice relaxation time of about a millisecond [34], which is a typical character of SMMs [6]. However, the studies on the growth and properties of the metal acetylacetonates on the surface are still rather sparse [41]. According to the calculation in Chapter 2, the Kondo effect should not appear in Dy(acac)$_3$ on Cu(111) surface and the ground states of the system are stable. In Chapter 5 low temperature STM studies on Dy(acac)$_3$ as well as chromium acetylacetonate (Cr(acac)$_3$) are presented.

2 Theoretical background

2.1 Scanning tunneling microscopy

Since the invention by Gerd Binnig und Heinrich Rohrer in 1982 [42, 10, 11], scanning tunneling microscopy (STM) became one of the most powerful techniques to investigate nanoscopic structures in surface science in the last decades. Using an atomically sharp tip to approach the conducting sample and applying a bias voltage between them, STM can reveal topological information and electronic properties of the sample. STM is based on measuring the current that tunnels through the vacuum barrier separating the sample and the tip because of the tunneling effect. Due to the exponential dependence of the tunneling current on the width of vacuum barrier, STM can reach a spatial resolution below 100 fm. It enables physicist to observe and manipulate the individual atoms and molecules directly [12].

2.1.1 The tunneling effect

In the following the tunneling process is described by using the simplest model. The electron is considered as a motion of a plane wave, that travels towards a finite potential barrier with a width a. The potential is described by

$$V(x) = \begin{cases} 0 & x < 0, \\ U_0 & 0 < x < a, \\ 0 & x > a. \end{cases} \tag{2.1}$$

It is assumed that the energy of electron E is smaller than the barrier hight U_0. Since the plane wave is scattered by the barrier, the wave is partially reflected from the barrier and partially transmitted. By solving the Schrödinger equation for this system, the wave function can be described by

$$\psi(x) = \begin{cases} e^{ikx} + Re^{-ikx} & x < 0, \\ Ae^{k'x} + Be^{-k'x} & 0 < x < a, \\ Se^{ikx} & x > a, \end{cases} \tag{2.2}$$

where

$$k = \sqrt{2mE}/\hbar, \tag{2.3}$$

$$k' = \sqrt{2m(U_0 - E)}/\hbar, \tag{2.4}$$

are the wave vectors, m the mass, E the energy of the electrons and $\hbar$ the reduced Plank constant.

Considering the continuity of wave functions and their first derivatives at $x = 0$ and $x = a$, one obtains the transmission probability

$$T = \left[1 + \frac{4k^2k'^2}{(k^2 + k'^2)^2} \sinh^2 k'a\right]^{-1}. \tag{2.5}$$

In the case of a STM, $e^{-k'a} \ll 1$ is typically fulfilled. The above formula can then be simplified to

$$T \approx \frac{16k^2k'^2}{(k^2 + k'^2)^2} e^{-2k'a} = \frac{16E(U_0 - E)}{U_0^2} \exp\left[-\frac{2a}{\hbar}\sqrt{2m(U_0 - E)}\right]. \tag{2.6}$$

The transmission probability depends exponentially on the width of barrier, i.e. the tunneling current depends exponentially on the tip-sample separation. For a metallic sample, the typical valve of k' is about 1 Å^{-1} [12], that means even a sub-Ångstrom change of distance causes a significant response on tunneling current, when the tip is scanned along the surface.

2.1.2 Bardeen's approach and Tersoff-Hamann model

The power of STM techniques comes from the fact that the STM is sensitive to three-dimensional structure of the surface. The simplest model of the tunneling effect, the one-dimensional tunneling problem, which has been solved in Section 2.1.1, is useful to illustrate the basic phenomenon related to STM. However, it is insufficient to quantitatively explain the results of STM experiments.

A more accurate theory to describe the tunnelling process between two electrodes was developed by Bardeen [43]. It can be used for various systems in particular for STM.

When a potential U is applied between two electrodes, an electron from an electrode with state μ and energy E_μ can tunnel to an unoccupied state ν of the other electrode with the energy $\nu + eU$. The tunneling probability is determined by the tunneling matrix element $M_{\nu\mu}$.

Using the first-order perturbation theory, the tunneling current flowing between the two electrodes is given by:

$$I = \frac{2\pi e}{\hbar} \sum_{\mu\nu} f(E_\mu)[1 - f(E_\nu + eU)]|M_{\nu\mu}|^2 \delta(E_\mu - E_\nu - eU), \tag{2.7}$$

where $f(E)$ is the Fermi function.

In Bardenn's work it has been shown that the tunneling matrix element $M_{\nu\mu}$ can be written down as:

$$M_{\nu\mu} = \frac{\hbar^2}{2m} \int (\psi_\mu^* \vec{\nabla} \psi_\nu + \psi_\nu^* \vec{\nabla} \psi_\mu) d\vec{S}, \tag{2.8}$$

where ψ_ν and ψ_μ are wave functions of the electrodes, i.e. tip and sample, in the absence of tunneling.

A further calculation of $M_{\nu\mu}$ needs a fixed electrode geometry. A typical STM geometry is shown in Fig. 2.1. The tip is modeled with a radius of curvature R. This model was first used by Tersoff and Hamman [12, 44, 45]. They assumed that only the s-wave solution is important. Under such assumptions, the tunneling current can analytically be calculated under the limit of small voltages and low temperature:

$$I \propto \rho_{tip}(E_F)\rho_{sample}(r_0, E_F) U e^{2R\sqrt{2m\phi}/\hbar}, \tag{2.9}$$

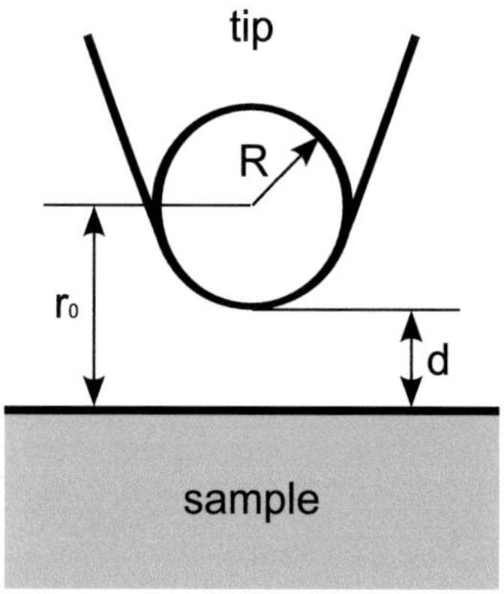

Figure 2.1: Scheme of the STM geometry used in the Tersoff-Hamann model. The tip is modeled with a radius of curvature R. d is the tip-sample separation.

where $\rho_{tip}(\varepsilon)$ is the density of states of tip, $\rho_{sample}(r_0, E_F)$ is the local density of stats (LDOS) of the sample at the centre of the tip sphere and ϕ is the work function. The LDOS of the sample can be written as a combination of the DOS of the sample $\rho_{sample}(\varepsilon)$ and the transmission coefficient $T(E,U,r)$, which only depends on the properties of the tunneling barrier:

$$\rho_{sample}(r_0, \varepsilon) = \rho_{sample}(\varepsilon) T(E,U,r_0). \tag{2.10}$$

For a larger bias, all the states that contribute to the tunneling current must be counted

$$I \propto \int_0^{eU} \rho_{tip}(E_F + \varepsilon - eU) \rho_{sample}(E_F + \varepsilon) T(\varepsilon, U, r) \mathrm{d}\varepsilon. \tag{2.11}$$

Using the WKB approximation [46], the transmission can be calcualted as

$$T(\varepsilon, U, r) \approx \exp\left\{ -2r \sqrt{\frac{2m}{\hbar}\left(\phi + \frac{eU}{2} - \varepsilon\right)} \right\}. \tag{2.12}$$

Equation (2.11) shows that the tunneling current actually represents the properties of LDOS of the sample rather than the topological structures. The DOS of tip has also a significant influence on the tunneling current. Thus, the tip is usually made of a malarial with a flat DOS near the Fermi edge, such as tungsten.

2.1.3 STM operation modes

The reverse piezoelectric effect allows to position an object with the accuracy of picometers by applying an electrical field, which causes a mechanical stain of piezoelectric ceramics. This enables the STM to scan along the surface with an atomically sharp tip carried by a piezotube.

The native operation mode of the STM is the so-called *constant height mode*. The tip scans over the surface by changing x-y position while keeping the z position (height) constant. By recording the tunneling current, information on the LDOS as well as on the topography of the sample is obtained.

However, as discussed in the previous sections, the tunneling current is extremely sensitive to the tip-sample separation. Even a height change of 3 Å, for example a monolayer step on the surface, causes a change of tunneling current by 3 orders of magnitude. Thus it is inconvenient to perform STM measurements using constant height mode. Furthermore, keeping the tip at constant height could cause unexpected tip-crashes with the surface. Thus an operation mode so-called *constant current mode* was developed. In this mode, the tunneling current is kept constant by using a feed-back loop, which adjusts the height of the tip according the change of the tunneling current during scanning.

The maximal moving range for one piece of piezo-material is typically hundreds of nanometers. This length is enough for the tip to scan a surface but insufficient for the tip to approach a fresh prepared sample. Pan *et al.* [47] designed an approach mechanism, in which several piezo-motors work collaboratively (see Fig. 2.2). The voltage is applied to the piezo-motors sequentially and each of them slips because of the frication between the other piezo-motors and the tip/sample. Then the voltage is withdrawn at the same time, so that all the piezo-motors move together and the tip/sample is moved one step. By repeating this process, a tip/sample can be moved by the piezo-motors in a range of several millimeters even at temperatures lower than 1 K.

2.1.4 Tunneling spectroscopy

As disscussed in Section 2.1.2, the tunneling current depends on the LDOS of a sample, which allows to measure the LDOS using STM.

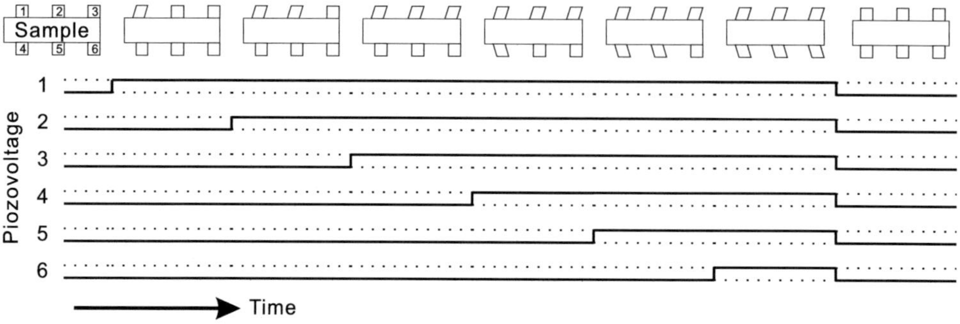

Figure 2.2: Scheme of the approach mechanism designed by Pan *et al* [47]. The sample/tip is moved one step by six piezo-motors working sequentially. By repeating this process, the sample can be moved millimeters.

Principles of tunneling spectroscopy

Starting from the results of the Tersoff-Hamann model (2.11), the first derivative of the tunneling current can be calculated:

$$\begin{aligned}\frac{\mathrm{d}I}{\mathrm{d}U} \propto & \frac{\mathrm{d}}{\mathrm{d}U}\int_0^{eU} \rho_{tip}(E_F+\varepsilon-eU)\rho_{sample}(E_F+\varepsilon)T(\varepsilon,U,r)\mathrm{d}\varepsilon \\ =&\rho_{tip}(E_F)\rho_{sample}(E_F+eU)T(eU,U,r) \\ &+\int_0^{eU}\frac{\mathrm{d}}{\mathrm{d}U}\rho_{tip}(E_F+\varepsilon-eU)\rho_{sample}(E_F+\varepsilon)T(\varepsilon,U,r)\mathrm{d}\varepsilon \\ &+\int_0^{eU}\rho_{tip}(E_F+\varepsilon-eU)\rho_{sample}(E_F+\varepsilon)\frac{\mathrm{d}}{\mathrm{d}U}T(\varepsilon,U,r)\mathrm{d}\varepsilon.\end{aligned} \tag{2.13}$$

Normally, it is reasonable to assume that $\rho_{tip}(E)$ and $T(\varepsilon,U,r)$ have no notable change near the Fermi edge compared to $\rho_{sample}(E)$. Thus the last two terms of (2.13) can be neglected and one obtains:

$$\frac{\mathrm{d}I}{\mathrm{d}U} \propto \rho_{sample}(E_F+eU). \tag{2.14}$$

This means the LDOS of sample is directly given by the first derivative of the tunneling current.

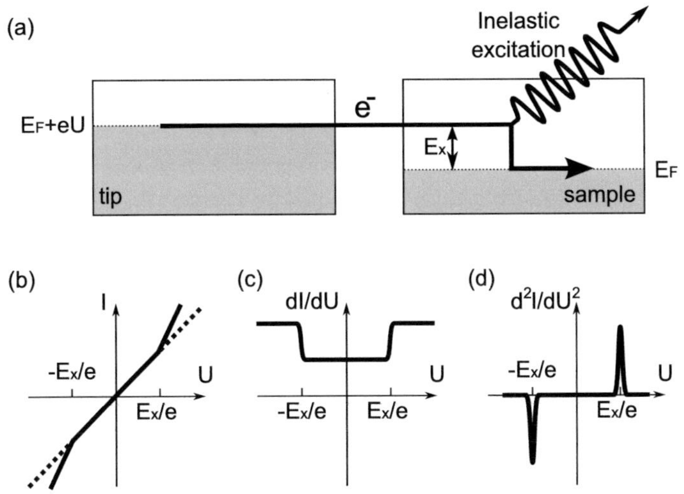

Figure 2.3: (a) Scheme of inelastic tunneling. (b) $I(U)$ curve of ITS. Due to the opening of the additional tunneling channel, $I(U)$ curve shows a larger slope above the excitation energy E_x. (c) Inelastic excitation appears as steps on the dI/dU curve of ITS. (d) d^2I/dU^2 curve of ITS. The anti-symmetric dip and peak indicate the inelastic excitation.

Inelastic tunneling spectroscopy (ITS)

So far, only the elastic tunneling process was discussed in which the energy of electron is conserved. However, the tunneling electron can also lose energy during the tunneling process due to interactions between the electron and the sample or the tip. A variety of excitations, like phonos [48, 49, 50, 51], plasmons [50] and magnons [28], can appear in the inelastic tunneling processes (see Fig. 2.3).

When the tunneling bias is small, only elastic tunneling can happen. As soon as the bias voltage reaches the value E_x/e, where E_x is the energy cost of the excitation, a second (inelastic) tunneling channel opens, which increases the conductivity of the tunneling junction. Thus, a kink appears in the $I(U)$ curve of ITS at the position $U = E_x/e$ (see Fig. 2.3(b)), while a step (peak) appears in the dI/dU (d^2I/dU^2) curve at the same position (see Fig. 2.3(c), (d)).

Compared with the elastic channel, the contribution of inelastic channel in the tunneling current is usually small. It could be difficult to directly recognize the kinks caused by inelastic excitations in the $I(U)$ curve. Therefore, it is advantageous to measure dI/dU or

d^2I/dU^2 curves, in which the signals of inelastic excitations, i.e. the steps in dI/dU curve and the anti-symmetric dips and peaks in d^2I/dU^2 curve, are more obvious.

Theoretically, the first and the second derivative can be numerically calculated directly from the $I(U)$ curve, usually the curves obtained by numerical derivation are rather noisy. Thus, in practice the dI/dU and d^2I/dU^2 curves are obtained by a lock-in amplifier, which modulates the input signal (bias) with a certain frequency and extract the Fourier component of the output signal (tunneling current) of the same frequency by multiplying it with the reference signal.

2.2 Magnetism

2.2.1 Spins in crystal field

Due to the exchange and the spin-orbit interaction, the electronic configuration of ground state of a multi-electron atom in free space follows the Hund's rules. The term with the lowest energy has a maximal total spin angular momentum S as well as the largest value of total orbital angular momentum L.

When the metallic atoms or ions interact with surrounding electrons in molecules or in a crystal via the Coulomb interaction, their electronic and magnetic properties change significantly. Because of the break of spherical symmetry, the orbital angular momentum L is not a good quantum number any more and the degeneracy of orbital electronic states is lifted, especially for d and f electrons. The orbital magnetic moment is (partially) quenched.

The so-called crystal field theory is a simplified model that describes the magnetic atom or ion in a molecule e.g. deposited on a metallic surface. The spin Hamiltonian can be written as [31]:

$$\hat{\mathscr{H}}_{spin} = \hat{\mathscr{H}}_Z + \hat{\mathscr{H}}_{CF}, \tag{2.15}$$

where $\hat{\mathscr{H}}_Z$ is the Zeeman Hamiltonian and $\hat{\mathscr{H}}_{CF}$ the crystal field Hamiltonian. The Zeeman Hamiltonian $\hat{\mathscr{H}}_Z$ describes the interaction between an applied magnetic field $\mathbf{H}$ and the magnetic moment $\mathbf{m}$ caused by spin $\mathbf{S}$:

$$\hat{\mathscr{H}}_Z = -\mathbf{H} \cdot \mathbf{m}, \tag{2.16}$$

where

$$\mathbf{m} = \mu_B g \cdot \mathbf{S}, \tag{2.17}$$

μ_B is the Bohr magneton and the g-factor is a tensor.

As a good approximation, the second term in 2.15, the crystal field Hamiltonian $\hat{\mathscr{H}}_{CF}$ can be written as [31]:

$$\hat{\mathscr{H}}_{CF} = DS_z^2 + E(S_x^2 - S_y^2). \tag{2.18}$$

Here $D$is the uniaxial magnetic anisotropy along the z-axis, which is normally vertical to the substrate surface. The second term represent the anisotropy in the x-y-plane, which is usually parallel to the surface of substrate. Equation (2.18) shows the break of the symmetry in the x-y plane due to crystal field. Even with the absence of a magnetic field $\mathbf{H}$, i.e. $\hat{\mathscr{H}}_Z = 0$, the spin levels are split by the crystal filed hamiltonian $\hat{\mathscr{H}}_{CF}$. This effect is often call zero-field splitting.

Considering the simplest situation, no magnetic field $\mathbf{H}$ and anisotropy in x-y-plane vanished, i.e. $E = 0$, only the uniaxial magnetic anisotropy $\hat{\mathscr{H}}_{spin} = DS_z^2$ is left. The system has totally $2S+1$ states: $|S_z> = |S>, |S-1>, ..., |-S+1>, |-S>$. Assuming an out-of-plane easy axis, $D < 0$, the system have two degenerate ground states: $|S>$ and $|-S>$ separated by the barrier with magnetic anisotropy energy (MAE) of $K = DS^2$ (see Fig. 2.4(a)).

By using ITS, the MAE can be evaluated. When tunneling electrons-carrying a spin $|\mathbf{S}_{el}| = 1/2$-tunnel to the magnetic atoms/molecules deposited on a metallic substrate, the inelastic scattering due to the exchange interaction between the spin of the tunneling electrons and the localized spin of the magnetic atoms/ions of the molecule can result spin-flip processes [29]. The z component $S_{z,el}$ of the spin of the tunneling electron can flip from $S_{z,el} = 1/2$ ($S_{z,el} = -1/2$) to $S_{z,el} = -1/2$ ($S_{z,el} = 1/2$) with $\Delta S_{z,el} = -1$ ($\Delta S_{z,el} = 1$). Due to the conservation of angular momentum, the magnetic atom/ion undergoes a spin transition with $\Delta S_{z,ion} = 1$ ($\Delta S_{z,ion} = -1$). When the invested system is at low temperatures, only the

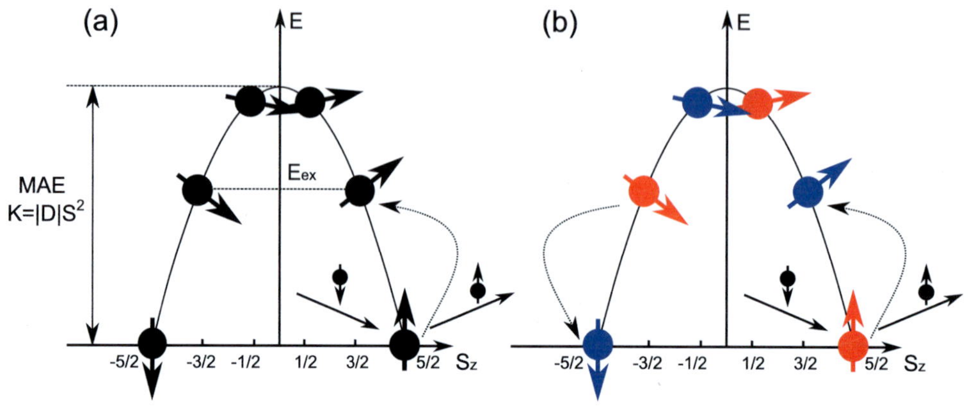

Figure 2.4: Energy scheme of a magnetic system with a spin of $|\mathbf{S}| = 5/2$. (a) In a crystal filed with a uniaxial MAE with $D < 0$. The two degenerate ground states: $|S_z = \frac{5}{2}>$ and $|S_z = -\frac{5}{2}>$ are separated by the barrier with MAE of $K = |D|S^2$. By inelastic spin-flip scattering the system at the ground state $|S_z = 5/2>$ ($|S_z = -5/2>$) can be excited to the higher state $|S_z = 3/2>$ ($|S_z = -5/2>$) with energy E_{ex}. (b) In a crystal filed with a non-zero $B_N^2\mathbf{O}_N^2$ term. The states $|S_z = 3/2>$, $|S_z = -1/2>$, $|S_z = -5/2>$ ($|S_z = -3/2>$, $|S_z = 1/2>$, $|S_z = 5/2>$) are mixed. In a spin-flip process the state $|S_z = 5/2>$ can tunnel into the state $|S_z = -5/2>$ via $|S_z = -3/2>$ and a Kondo resonance can arise.

ground states $|S_{z,ion} = S>$ and $|S_{z,ion} = -S>$ are occupied. Therefore, the magnetic particle at the ground state $|S_{z,ion} = -S>$ ($|S_{z,ion} = S>$) is excited to a higher state $|S_{z,ion} = -S+1>$ ($|S_{z,ion} = S-1>$).

Considering $\hat{\mathscr{H}} = DS_z^2$, the energy cost in this precess is

$$E_{ex} = D(S-1)^2 - DS^2 = |D|(2S-1). \tag{2.19}$$

Thus, only the tunneling electron with an energy $E > E_{ex} = |D|(2S-1)$ can cause spin-flip excitation. This results in a dip-peak signal at $V_{ex} = \pm E_{ex}/e$ on ITS (see Section 2.1.4). So the MAE can be evaluated experimentally

$$\mathrm{MAE} = |D|S^2 = \frac{S^2}{2S-1} eU_{ex}. \tag{2.20}$$

In the above discussions it was assumed that the orbital angular momentum $\mathbf{L}$ is totally quenched due to crystal filed and only spin $\mathbf{S}$ presents. In the case of a partial quench of $\mathbf{L}$, the spin-orbit coupling must be considered and the spin $\mathbf{S}$ ((2.17)-(2.20)) must be replaced by $\mathbf{J} = \mathbf{S} + \mathbf{L}$ [52].

The Hamiltonian $\hat{\mathscr{H}}_{CF}$ (see Equation (2.18)) is the simplest case of a crystal field Hamiltonian. In principle, $\hat{\mathscr{H}}_{CF}$ could contain higher order terms which can be developed as the sum of the operators of angular momentum [31]:

$$\hat{\mathscr{H}}_{CF} = \sum_{N,k} B_N^k \mathbf{O}_N^k, \tag{2.21}$$

where B_N^k are parameters and $\mathbf{O}_N^k$ are the so-called Steven operators, which can be expressed as a function of operators of the type $\mathbf{J}_x^k$, $\mathbf{J}_y^k$, $\mathbf{J}_z^k$ and the integer k satisfies

$$-N \leq k \leq +N. \tag{2.22}$$

Due to the time reverseal symmetry, N must be even and considering that the higher terms are comparatively smaller than the lower order terms, it is only necessary to include the terms of order

$$N = 2, 4, 6, ..., 2J. \tag{2.23}$$

The $k = 0$ term depends only on the operators $\mathbf{J}_z^N$, therefore it is diagonal in the J manifold and changes the energy levels of the the $|J_z>$ states. The $k \neq 0$ terms, howsoever, contains the operators $\mathbf{J}_x^k$ and $\mathbf{J}_y^k$ and couples the states differing by $\pm k$ in m, so that the $|J_z>$ states are not eigenstates any more. The new eigenstates are the linear combinations of different $|J_z>$ (see Fig. 2.4(b)):

$$\begin{aligned}
|\psi_1> &= \alpha_1|J> + \alpha_2|J-k> + \alpha_3|J-2k> + ... \\
|\psi_2> &= \beta_1|J-1> + \beta_2|J-1-k> + \beta_3|J-1-2k> + ... \\
. &\quad . \\
. &\quad . \\
|\psi_{2J+1}> &= \gamma_1|-J> + \gamma_2|-J+k> + \gamma_3|-J+2k> + ...
\end{aligned} \tag{2.24}$$

In this case, to keep a stable magnetic moment in the crystal field, the two states $|J>$, $|-J>$ of magnetic atom/ion should not mix, which means the states $|-J>$ should not appear in Equation (2.24). In other words $|-J> \neq |J-mk>$, $m \in \mathbb{N}$, or

$$\frac{2J}{k} \neq m, m \in \mathbb{N}. \tag{2.25}$$

On the other hand, considering that $|-J+1>$ can tunnel into the sates $|-J>$ in a spin-flip process with $\Delta J_z = 1$, if the states $|J>$ was mixed with the states $|-J+1>$ in Equation (2.24), the spin-flip process can couple the two states $|J>$ and $|-J>$ without paying additional energy. In this case, the Kondo effect can arise in the system (see Section 2.2.3). To avoid the Kondo effect, following equation must be satisfied:

$$\frac{2J-1}{k} \neq m, m \in \mathbb{N}. \tag{2.26}$$

So, the Equations (2.25) and (2.26) are the conditions to keep a stable magnetic moment of a magnetic atom/ion in the crystal field in tunneling process. Keeping in mind that these conditions must be satisfied with every non-zero k appearing in the parameters B_N^k of $\mathscr{H}_{CF}$ (see Equation (2.21)).

However, in a system with a certain point group symmetry, some terms in (2.21) vanish ($B_N^k = 0$). For instance, considering a crystal field with a threefold symmetry C_3, it can be proven that only the terms of B_N^0, B_N^3, B_N^6 appear in the crystal field Hamiltonian, which contain only the operators $\mathbf{J}_+^3$, $\mathbf{J}_-^3$ and $\mathbf{J}_z$ [31, 53]. Thus the states $|J_z = m>$, $|J_z = m \pm 3>$, $|J_z = m \pm 6>$, ... are mixed. For a Dy(III) ion (see Chapter 5) with the electron configuration $4f_9$, the total angular moment is $J = 15/2$. So the condition $\frac{2J-1}{k} \neq m$, $m \in \mathbb{N}$ is fulfilled with $k = 3$. Therefore a Kondo resonance should not arise when a Dy(III) ion locates in a crystal field with a threefold symmetry. For a Cr(III) ion, the electron configuration is $3d_3$. Because the Cr is transition metal, one can assume that the orbital angular momentum $\mathbf{L}$ totally quenches in crystal field and $\mathbf{J} = \mathbf{S} = 3/2$. So $\frac{2J-1}{k} = 1$, when $k = 2$, and one can predict that a Kondo resonance will appear when a Cr(III) ion locates in a crystal field of twofold symmetry.

2.2.2 Antiferromagnetically coupled dimers

So far, the interaction between localized spins with the environment is discussed using a simplified crystal field model. The spin of magnetic atoms/molecules was treated as a whole. However, magnetic molecules often contain more than one magnetic center coupled ferromatically or antiferromatically via exchange interactions [31, 54, 6]. The internal spin configurations of magnetic molecules could be influenced by the inelastic scattering with tunneling electrons.

The interaction inside a magnetic system with a group of localized magnetic momenta can be described by the Heisenberg-model:

$$\hat{\mathscr{H}} = \sum_{\mu<\nu} J_{ex,\mu\nu}\hat{s}_\mu \cdot \hat{s}_\nu, \tag{2.27}$$

where $J_{ex,\mu\nu}$ is the strength of the coupling between two magnetic centers with spin s_μ and s_ν respectively.

A general solution to (2.27) does not exist. It can, however, be solved analytically for the simplest system containing only two spins, i.e. a Hamiltonian that can be written down as

$$\begin{aligned} \hat{\mathscr{H}} &= -J_{ex}\hat{s}_1 \cdot \hat{s}_2 = -\frac{1}{2}J_{ex}(\hat{S}^2 - \hat{s}_1^2 - \hat{s}_2^2) \\ &= -\frac{S(S+1) - s_1(s_1+1) - s_2(s_2+1)}{2}J_{ex}, \end{aligned} \tag{2.28}$$

where $\hat{S} = \hat{s_1} + \hat{s_2}$ and S may has the value $|s_1 - s_2|$, $|s_1 - s_2| + 1$, ..., $s_1 + s_2 - 1$, $s_1 + s_2$. When the two spins couple ferromagnetically, i.e. $J_{ex} > 0$, the energy is lower by arranging the two spin parallely. Therefore, the ground state has total spin $S = s_1 + s_2$. When the two spins couple antiferromagnetically, i.e.$J_{ex} < 0$, the energetic minimum is achieved by arranging the two spin antiparallelly and the ground state has total spin $S = |s_1 - s_2|$.

Now we consider a system with two magnetic centers with spin $s_1 = s_2 = s$ and an antiferromagnetic coupling $J_{ex} < 0$. The ground states is $S = 0$, which is the superposition of $|m_{s_1} = s, m_{s_2} = -s>$, $|m_{s_1} = s-1, m_{s_2} = -s+1>$, ..., $|m_{s_1} = -s+1, m_{s_2} = s-1>$ and $|m_{s_1} = -s, m_{s_2} = s>$. The superposition coefficients are the Clebsch-Gordan coefficients.

Assuming the system is at a low temperature, i.e. $|J_{ex}| >> k_B T$, only the ground state is occupied.

When the system is investigated using STM, inelastic scattering with tunneling electrons can result a spin-flip excitation of one magnetic center. The z-component of the tunnel electron flips from $m_{s,el} = 1/2$ ($m_{s,el} = -1/2$) to $m_{s,el} = -1/2$ ($m_{s,el} = 1/2$) with $\Delta m_{s,el} = \mp 1$ while the spin of the magnetic center flips from $|m_{s_1}, m_{s_2}>$ to $|m_{s_1}+1, m_{s_2}>$ ($|m_{s_1}-1, m_{s_2}>$) or $|m_{s_1}, m_{s_2}+1>$ ($|m_{s_1}, m_{s_2}-1>$). This means the system is excited from the ground state $S = 0$ to one of the three states of the triplet in $S = 1$ with the selection rule of $\Delta m_S = \pm 1$.

By using (2.28) one obtains the excitation energy of this spin-flip process:

$$\begin{aligned} E_{ex} &= -\frac{1(1+1) - s_1(s_1+1) - s_2(s_2+1)}{2} J_{ex} \\ &\quad + \frac{0(0+1) - s_1(s_1+1) - s_2(s_2+1)}{2} J_{ex} \\ &= -J_{ex}, \end{aligned} \tag{2.29}$$

which can be experimentally determined by performing ITS (see Section 2.1.4 and 2.2.1).

2.2.3 The Kondo effect

As one of the important many-body phenomena in condensed matter physics, the Kondo effect has attracted a lot of attention from both theorists and experimentalists.

Originally, the Kondo effect describes a phenomenon that was found in the 1930s [55]. Below about 10 K, the electrical resistance of metals with magnetic impurities, such as cobalt, increases as the temperature is lowered further. In 1964 this phenomenon was explained by Kondo for the first time [56] by considering the scattering of electrons by localized magnetic impurities. However, Kondo's theory predicted an infinite resistance at even lower temperatures. In the late 1960s, Anderson's model [57] was used in the theoretical framework to explain the properties of real systems close to 0 K. In the 1970s, the renormalization group theory was introduced to explain the Kondo effect [9]. It has been shown that at low temperatures, the magnetic moment of the impurity is screened by surrounding electrons of the substrates.

Figure 2.5 shows a simple model introduced by Anderson in 1961 to explain the Kondo effect [58]. It is assumed that the magnetic impurity has only one electron level E_0 below

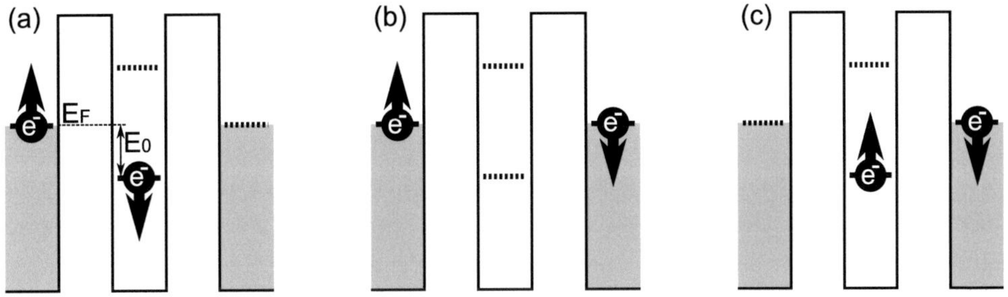

Figure 2.5: Scheme of the Anderson model [58]. (a) The initial state. The magnetic impurity has one electron level E_0 below the Fermi level E_F. (b) The virtual intermediate state. The spin-up electron tunneled out of the impurity and occupied a classically forbidden virtual state. (c) The final state. The unoccupied impurity state has been filled by an electron with spin-down from the Fermi sea of the metal. The initial and final states of the election of the magnetic impurity have opposite spins.

Fermi level E_F. Due to the Heisenberg uncertainty principle, the electron can jump out from the localized impurity state and take an unoccupied state at the Fermi level of the metal for a short time $\sim h/E_0$, where h is the Planck constant. At the same time, another electron from the Fermi sea with opposite spin can tunnel into the unoccupied impurity state E_0. Finally, the initial and final states of the electron of the magnetic impurity have reversed spins.

After the 1990s, the Konto effect was observed in transport measurements across quantum dots [59, 60, 58] as well as in STM measurements on magnetic adatoms [61, 62, 63, 64, 65, 66] and metallic molecules with magnetic ions [67, 64, 68]. In electron transport across the quantum dots or single atoms and molecules, the Kondo resonance provides additional density of states near the Fermi level, which appears as a zero bias abnormality in the spectra (STS).

As the simplest manifestation of the interaction between localized electrons with delocalized electrons, the shape of the zero bias abnormality in STS caused by the Konto effect can be described by a Fano resonance [69, 70, 71]:

$$\frac{dI}{dU}(U) \propto \frac{(\varepsilon + q)^2}{1 + \varepsilon^2}, \tag{2.30}$$

where

$$\varepsilon = \frac{eU - \varepsilon_0}{\Gamma}, \tag{2.31}$$

and ε_0 is the energy shift of the resonance from the Fermi level, Γ is the half width of the resonance. The Fano parameter q characterizes the interference of the tunneling between the tip and the magnetic impurity and the tunneling between the tip and the sample. It is defined by [72]

$$q \equiv \frac{t_a}{2\pi V t_c}, \tag{2.32}$$

where t_a is the matrix element of interaction between the probe (tip) and the localized (spin) state, t_c is the matrix element of coupling between the probe (tip) and continuum (sample) and V the the hybridization matrix element between the localized state and the continuum. Furthermore, q determines the shape of the Fano resonance (see Fig. 2.6). Considering the temperature dependence, the Kondo resonance can be approximated by a Lorentzian resonance [73]. Thus the energy width 2Γ (FWHM) of a Kondo peak can be expressed as:

$$\Gamma = 2\sqrt{(\pi k_B T)^2 + 2(k_B T_K)^2}, \tag{2.33}$$

where k_B is the Boltzmann constant, T the environment temperature, T_K the Kondo temperature. According to numerical renormalization group calculation [74, 75], the temperature dependence of the maximum intensity of the Fano resonance (G_K), which appears at the Fermi level, can be expressed as:

$$G_K(T) = G(0)\left[1 + \left(\frac{T}{T_K}\right)^2 \left(2^{1/\alpha} - 1\right)\right]^{-\alpha}, \tag{2.34}$$

where $G(0)$ is the intensity at $T = 0$, and α a parameters that depend on the angular momentum of the impurity.

The characteristic asymmetric peak near the Fermi level in STS can be used as an evidence of the appearance of Kondo resonance in the investigated system.

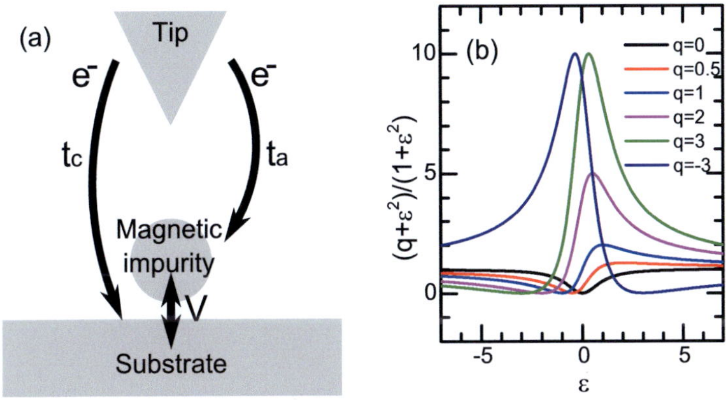

Figure 2.6: (a) The Fano parameter q characterizes the interference of the tunneling between the tip and the magnetic impurity and the tunneling between the tip and the sample. (b) Line shapes of Fano resonance for different values of q [69].

3 Experimental setup

For the study of nanoscopic structures, especially single molecular magnets (SMMs), an instrument with high lateral resolution and high energy resolution for electronic states is needed. A usual solution for this requirement is a scanning tunneling microscope (STM) [10, 11, 12] working at cryogenic temperatures [13, 14, 15, 16, 17, 18, 19, 20]. For surface science studies of atomically clean structures, the STM needs to be operated in ultra high vacuum (UHV) [12, 21, 22, 23]. Typical commercially available low temperature STMs working in UHV use normally He-bath cryostat and operate at $\geq$4.2 K [24, 25, 26]. The energy resolution of inelastic tunneling spectroscopy (ITS) at the temperature T is limited to 5.4 $k_{B}T$ [27] which is only about 1.9 meV at 4.2 K. For many interesting questions such as magnetic excitations of single atoms and molecules [28, 29, 30], this resolution is insufficient and lower temperatures are required. For many studies on these magnetic nanostructures, the use of magnetic tips is advantageous [76], requiring frequent tip preparation with easy tip exchange entirely in UHV. To meet all these requirements, an ultra stable and low noise STM operating at 400 mK using ^{3}He (930 mK using ^{4}He) has been developed [77].

3.1 Technical properties

The entirely homebuild STM including a ^{3}He cryosystem was developed and built. Special care was taken that the consumption of cryogenic liquids stays low and that the vibrational level of the instrument at the STM tip is in the range of several 100 fm, only.

A three stage cryostat is used to cool down the microscope head. The first stage is cooled with liquid nitrogen, the second stage with liquid ^{4}He. The third stage uses a closed-cycle Joule-Thomson refrigerator of a cooling power of 1 mW. A base temperature of 930 mK at the microscope head was achieved using expansion of ^{4}He, which can be reduced to $\approx$ 400 mK when using ^{3}He. The cryostat has a low liquid helium consumption of only 38 ml per hour and standing times of up to 280 hours. The fast cooling down of the samples (3 h) guarantees high sample throughput. Test experiments with a superconducting tip show a high energy resolution of 0.3 meV when performing scanning tunneling spectroscopy. The vertical stability of the tunnel junction is well below 1 pm (peak to peak) and the electric noise floor of tunneling current is about 6 $fA/\sqrt{Hz}$. Atomic resolution with a tunneling current of 1 pA and 1 mV was achieved on Au(111). The lateral drift of the microscope at stable temperature is below 20 pm/h. A superconducting spilt-coil magnet allows to apply an out-of-plane magnetic field of up to 4 T at the sample surface. The flux vortices of a Nb(110) sample were clearly resolved in a map of differential conductance at 1.1K and a magnetic field of 0.21 T. The set-up is designed for in-situ preparation of tips and samples under UHV condition.

3.2 System design

Figure 3.1 shows the overall design of the UHV recipient including the cryostat chamber (1) and the preparation chamber (2). The overall height is kept low to allow assembly and operation of the set up in rooms of only 3 m height. The microscope head (6) is placed inside the cryostat chamber, while tip and sample preparation are carried out in the preparation chamber.

3.2.1 Cryostat

The development of the compact cryostat was one of the main parts of this work. The overall aim was to achieve sub-kelvin temperatures at the microscope head even without using expensive ^{3}He. To minimize the consumption of liquid He and liquid N_2, the input heating power on the cryogenic baths must be as low as possible. For quick sample and tip exchange, easy access to the microscope head and good thermal conductivity between the microscope head and the cryostat are both necessary. Furthermore, the microscope head must be isolated from the mechanical vibration from the cryogenic liquids. To achieve UHV conditions, it is indispensable that the system can be baked out at high temperatures (150 $\sim$ 250 °C). For all of these requirements, a home-built cryostat using the Joule-Thomson effect was developed.

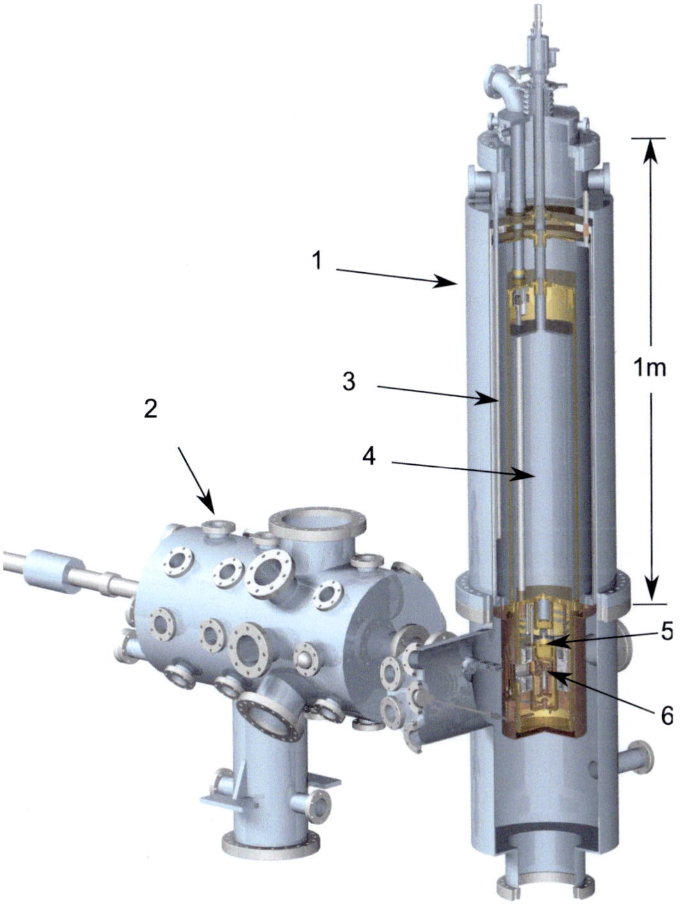

Figure 3.1: Layout of the UHV recipient with (1) the cryostat chamber, (2) the preparation chamber, (3) liquid nitrogen bath cryostat, (4) liquid helium bath cryostat, (5) Joule-Thomson cryostat, (6) microscope head.

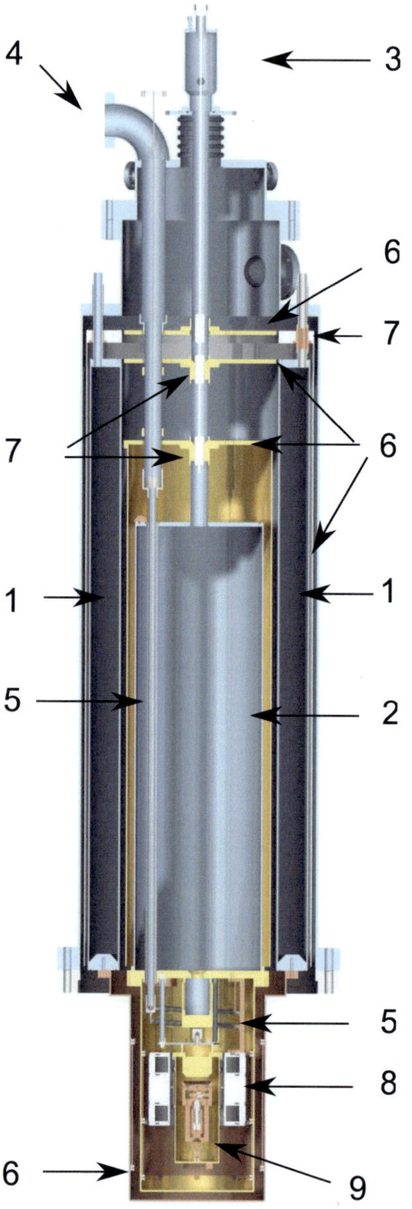

Figure 3.2: Scheme of the cryostat. (1) liquid nitrogen tank, (2) liquid helium tank, (3) in and out let of liquid helium, (4) in and out let of ^{3}He, (5) countercurrent heat exchanger for ^{3}He, (6) radiation shield, (7) heat exchangers coupled with cryogenic gas (8) superconductive coil (9) microscope head.

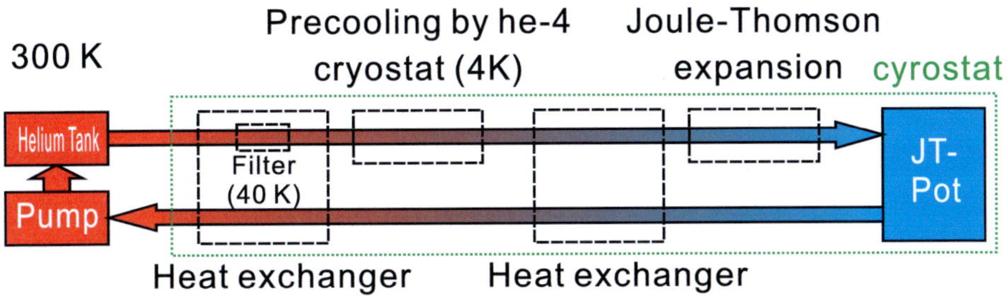

Figure 3.3: Schematics of the ^{3}He cycle in Joule-Thomson cooling stage.

Design of the cryostat

Figure 3.2 shows a scheme of the cryostat with a total height of 120 cm and a diameter of 30 cm consisting of three cooling stages. The first and the second stage are concentric bath cryostats containing 20 liters liquid N_2 (1) and 10 liters liquid ^{4}He (2), respectively. All the containers of the cryogenic liquids are radiation shielded (6) so that the input thermal power remains low. The shields themselves are of high thermal conductivity (oxygen free copper) coated with a highly reflective material (silver or gold) to minimize the impact of heat radiation. Moreover, the shields around the nitrogen and helium baths are thermally coupled to the exhausting cryogenic gas via heat exchangers (7). Therefore, the heat capacity of the cold gas is used to further cool the radiation shields. As the main source of heat input on the cold bath is given by thermal radiation, which scales with T^4, even a small decrease of the temperature of the shields substantially lowers the consumption of the cryogenic liquids.

The last cooling stage consists of a ^{3}He Joule-Thomson cooler. Its working principle is shown in Fig. 3.3. ^{3}He gas is injected into the cryostat at room temperature and at $p <1.5$ bar and is subsequently pre-cooled in a first concentric counter-current heat exchanger mounted inside the ^{3}He pumping line. To prevent impurity in the ^{3}He gas that could block the Joule-Thomson expansion capillary, a filter with a volume of $\approx$ 1.3 cm^3, which is thermally anchored to the cooled radiation shield at $\approx$ 40 K, was inserted into the first heat exchanger.

In a next step, the ^{3}He is cooled down to 4.2 K within a one meter long pipe, which is thermally anchored to the bottom of the liquid helium bath cryostat. This is followed by a second countercurrent heat exchanger consisting of a 10 m spiral encapsulated in the pumping line. Finally, the gas can freely expand through a capillary with a diameter of 90 μm, thereby reaching sub-kelvin temperatures due to the Joule-Thomson effect. A reservoir of 10 ml is placed at the end of the capillary, which is pumped by an external pump to a pressure of ≈ 0.1 mbar. The total throughput of ^{3}He gas in steady state corresponds to 3 liters gas per hour under ambient conditions giving an estimated cooling power of 1 mW at 500 mK [78]. Below the Joule-Thomson cryostat, the cryogenic STM is mounted. It is thermally anchored to the Joule-Thomson stage by thin Au wires (50 μm). The input of heat is kept at a minimum (<100 μW) by thermal insulation of the ^{3}He stage from the ^{4}He cooled parts of the cryostat. This enables to use the ^{3}He bath also in the single shot mode.

As the ^{3}He stage is well insulated from the ^{4}He stage, cooling down of new samples and tips would be very slow. Thus, after inserting a sample good thermal coupling between microscope head and liquid helium bath cryostat is required. To overcome this dilemma, a mechanical thermal switch, which can be easily controlled from the outside, was integrated. By using this switch, it only takes three hours to cool down a sample and a tip from room temperature to sub-kelvin temperatures ensuring both quick tip and sample exchange and sub Kelvin operation.

The performance of the cryostat was modeled by linear heat transport equations taking heat flows due to radiation, heat conduction and the heat exchangers into considerations. As a good approximation, it was assumed that the containers of cryogenic liquids have a homogeneous temperature distribution as well as the radiation shields which were made of either aluminum or copper. The heat conduction inside cryogenic liquids and exhausting cryogenic gas was neglected. In this case, only the contribution of heat conduction between exhausting cryogenic gas and heat exchangers, and inside the pipes as in- and out-let of cryogenic liquids, were considered. The emissivity of polished metal surface at a wavelength of about 10 μm (corresponds to room temperature) was taken to calculate radiative heat transfer between parts of different temperature [79]. Then the efficiency of heat exchangers was treated as free parameters in the model. They were later estimated by fitting the model with the experimental consumption of cryogenic liquids.

Test of Joule-Thomson cryostat with ^{4}He

The system was manufactured, assembled and tested. Due to the lack of ^{3}He on the international market, the Joule-Thomson stage was operated with ^{4}He instead of ^{3}He at first. The careful design of the cryostat minimizes the input heat power to 8.5 W to the 77 K bath and 26.5 mW to the 4.2 K bath, so that the consumption of only 153 g/h of liquid N_2 and 4.7 g/h of liquid ^{4}He was achieved. This corresponds to standing times of 105 hours for

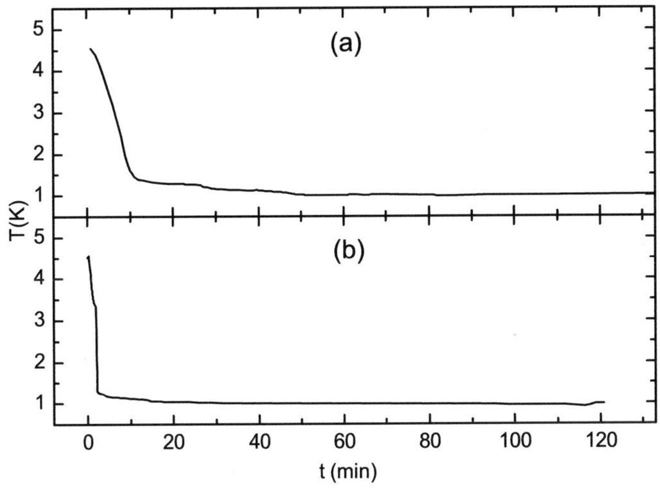

Figure 3.4: Temperature of (a) the STM head and (b) the Joule-Thomson cryostat as a function of time after starting the Joule-Thomson expansion of ^{4}He.

liquid nitrogen (20 liters) and of 280 hours for liquid helium (10 liters). Keeping in mind that the cryostat has to absorb the heat input of the ^{3}He injection and pumping lines (contributing approximately 50% of the total heat load on the 4.2 K bath) as well as that of the 50 A current lines for the superconducting magnet, this consumption is extraordinary low. It could only be achieved by the design of highly efficient heat exchangers (~90% efficiency for the nitrogen and ~70% for the helium) that use the heat capacity of the cold exhausting gas.

A calibrated Cernox cryogenic temperature sensor [80] was mounted on the Joule-Thomson stage and the microscope head to measure the temperature. In test experiments of the Joule-Thomson expansion using ^{4}He, the cryostat was cooled from 4.5 K to below 1 K in less than 20 minutes (see Fig. 3.4(b)). Similarly, the STM was quickly cooled down to about 1.1 K during filling the Joule-Thomson cryostat in the first 20 minutes, which indicates an effective thermal coupling. After stopping the He input, a base temperature of 930 mK was achieved until all the ^{4}He has been consumed (see Fig. 3.4(a)). In case the cavity of 10 ml in the last stage of the cryostat was completely filled with liquid ^{4}He and the Joule-Thomson expansion was stopped, standing times of the order of 20 hours at the base temperature could be achieved with an operating STM indicating the low heat input on the sub Kelvin components of the set up.

Because the vapor pressure above the helium surface depends mainly on the pumping speed of the Joule-Thomson exhaust system, it is reasonable to assume that the vapor pressure inside the Joule-Thomson cryostat is essentially independent on the isotope of He being used. Considering the helium vapor pressure as one of the most common scales of low temperatures [81, 82], the base temperature of 930 mK for ^{4}He converts to an expected base temperature of 420 mK when being operated with ^{3}He.

^{3}He handling system

After testing the Joule-Thomson stage of the cryostat with ^{4}He, a compact home-built ^{3}He handling system was manufactured that handles pumping, storing and purification of ^{3}He.

Figure 3.5 shows a scheme of the ^{3}He handling system. After manufacturing, the system was carefully checked for Helium leaks to make sure that the total leak rate was below 10^{-8} liters per seconds. This guarantees that the ^{3}He-loss will be neglectable in the system after running for 20 years. After evacuating, the system was feed with ^{3}He through valve-8 (v8), and ^{3}He was stored in the storage tank (c). The amount of gas was controlled by the pressure gauge (p1). The vacuum rotary pump (b) was used to generate a vacuum of ~ 0.1 mbar (controlled by pressure gauge p3) to pump the ^{3}He out of the cryostat (a), compress the gas to 1.5 bar (controlled by pressure gauge p2) and pump it back into the cryostat to realize the cycling. The pump is located in a sound-proof box (h) in order to prevent the sensitive STM-experiments being disturbed by the noise from the pump.

Because the Joule-Thomson stage of the cryostat works at a temperature less than 1 K, any impurity appearing in the ^{3}He freezes at this temperature and blocks the capillary of the cryostat. Several filters (d, e, f, g) were installed in the ^{3}He handling system to keep the gas clean. A filter made up of steel wool (e) is used to stop the oil evaporated from the pump, which condenses in the filter and drops back into the pump. An active carbon filter which is cooled by liquid nitrogen (d) can adsorb most of the other impurities. The impurities with a melting point above 77 K turn solid and stay in the filter. After being used in a hot environment (about 60 °C in this case) for some time, the pump oil deteriorates and emits some hydrogen, which can go though the 77 K active-carbon filter and blocks the capillary of the cryostat. To prevent this, a hydrogen filter made up of palladium powder (f) was installed. After a period of time, both of the hydrogen filter and 77 K active-carbon filter could be saturated. Then they are baked out at 200 °C, so that the adsorbed impurities desorb. An external filter cooled by liquid helium (g) removes them from the cycle.

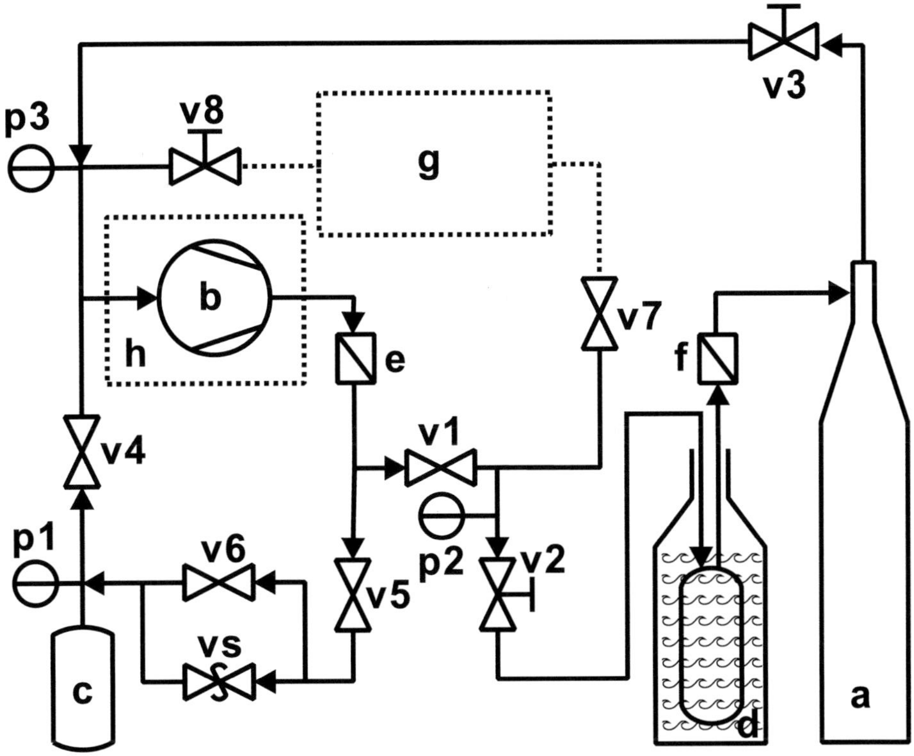

Figure 3.5: Scheme of the ^{3}He handling system. (a) cryostat, (b) rotary pump, (c) ^{3}He storage tank, (d) liquid nitrogen cooled active carbon filter, (e) pump oil filter made up of steel wool, (f) hydrogen filter made up of palladium powder, (g) external liquid helium filter for hydrogen, (h) sound-proof box, (p1)-(p3) pressure gauges, (v1)-(v9) valves. (vs) safety valve.

When the temperature of the Joule-Thomson stage of the cryostat increases due to a sample transfer, all the liquid ^{3}He stored in the cryostat turn gaseous in a short time. The pressure between rotary pump and cryostat increase rapidly in this case. As soon as the pressure exceeds 1.5 bar, a safety valve (vs) will release the excess gas back into the storage tank to prevent damage of the equipment as well as catastrophic loss of ^{3}He caused by too high overpressure.

The gas flowing in this ^{3}He handling system is controlled by operating the valves (v1-v8) manually. Table 3.1 shows the status of every valve during the different operating status of the system.

Table 3.1: Different operating status of ^{3}He handling system

operating status	v1	v2	v3	v4	v5	v6	v7	v8
evacuating ^{3}He handling system	open	open	open	open	open	open	open	closed
leak testing (system empty)	open	open	open	open	open	open	open	closed
feeding system with ^{3}He	open	closed	open	closed	open	open	open	closed
increasing pressure of JT-stage	open	open	open	open	open	closed	closed	closed
JT-stage running	open	open	open	closed	open	closed	closed	closed
reducing pressure of JT-stage	open	open	open	closed	open	open	closed	closed
JT-stage stop (pump running)	closed	open	open	closed	open	open	closed	closed
JT-stage stop (pump stop)	closed	open	closed	closed	open	open	closed	closed
filtering ^{3}He with extern LHe filter	open	closed	open	closed	open	closed	open	open

Test of Joule-Thomson cryostat with ^{3}He-^{4}He mixture

With the ^{3}He handling system, the Joule-Thomson stage of the cryostat can be operated theoretically with pure ^{3}He. However, before starting the operation of the Joule-Thomson cryostat, the temperature of the capillary for Joule-Thomson expansion cooled by the liquid Helium stage is about 5 K. Considering that the maximal injection pressure of ^{3}He is 1.5 bar, which is limited by the rotary pump, the Joule-Thomson coefficient μ_{JT} of ^{3}He [83], which is defined by

$$\mu_{JT} = \left(\frac{\partial T}{\partial p}\right)_H, \tag{3.1}$$

is negative at this situation [84, 83] (see Fig. 3.6). That means, the Joule-Tomson expansion warms up the ^{3}He instead of cooling down.

There are two possible methods to solve this problem. One of them is, before starting operating the Joule-Thomson stage, pumping the liquid helium bath-cryostat with a rotary pump and reducing the pressure to ≤ 0.1 bar so that the temperature of liquid helium cryostat as well as Joule-Thomson cryostat are cooled below 3 Kelvin. However, this methods causes additional time to cool down the sample and the rotary pump generates noise and disturbs the STM measurements.

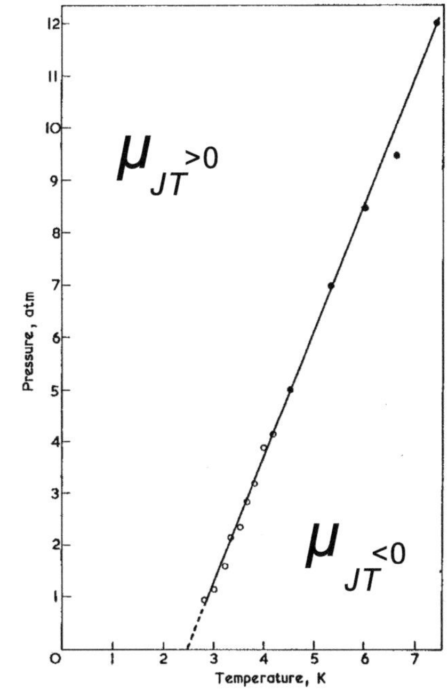

Figure 3.6: Inversion curve of Joule-Thomson effect of ^{3}He [84]. When pressure $P <$ 1.5 bar, temperature $T > 4$ K, the Joule-Thomson coefficient μ_{JT} of ^{3} is negative, which means Joule-Tomson expansion will increase the temperature of ^{3}He.

Another way to make Joule-Thomson stage functional is using a ^{3}He-^{4}He mixture instead of pure ^{3}He in the Joule-Thomson stage. When the Joule-Thomson expansion starts, the ^{4}He in the mixture, which has a positive μ_{JT} at $\sim$ 5 Kelvin and 1.5 bar, cools the ^{3}He part as well as the Joule-Thomson cryostat. When the temperature is below 1 Kelvin, the vapor pressure of ^{4}He is much lower than ^{3}He [78, 79]. Then, the ^{4}He concentrates as liquid in the reservoir of Joule-Thomson stage and ^{3}He in the mixture cools the system further.

In our case, the second method has been chosen. The Joule-Thomson stage of the cryostat was operated using a mixture of 0.5 gram ^{3}He and 0.6 gram ^{4}He instead of pure ^{3}He. With the cooling process shown in Fig. 3.7, a temperature of 650 mK at the STM head was achieved. Even lower temperatures are expected after optimization of the insertion pressure into Joule-Thomson stage and the ^{3}He-^{4}He ratio of the mixture.

Discussions about pumping power and cooling power of Joule-Thomson stage

During the testing the Joule-Thomson stage with pure ^{4}He as well as ^{3}He-^{4}He mixture, precise values of some physical properties of the cryostat that were difficult to calculate (such as Helium flux and total heat input), were experimentally obtained. This allows to calculate the temperature profile and pressure distribution inside the Joule-Thomson stage.

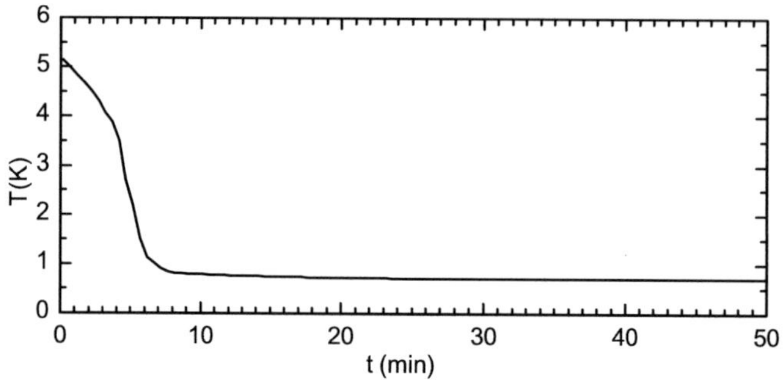

Figure 3.7: Temperature of the STM head as a function of time after starting the Joule-Thomson expansion of the ^{3}He-^{4}He mixture.

After stopping injecting ^{3}He-^{4}He mixture into the Joule-Thomson stage and after all liquid in the reservoir has been exhausted, the STM head warms up. By measuring the warming up rate at 2 K, the total heat input to the STM head and the Joule-Thomson stage was obtained, which is $9\mu W$ at 2 K. Considering the latent heat of vaporization of ^{3}He is $\sim$ 25 j/mol below 1 K [78, 79], the vaporization rate of liquid ^{3}He in the reservoir should be $\sim$ 0.4 μmol per second, after stopping injecting gaseous ^{3}He into cryostat.

However, the actual ^{3}He flow is measured to be $\sim$ 5 μmol per second, which is one order of magnitude larger than the predicted value. That means there is additional heat input or an additional vaporization process at the Joule-Thomson stage even after stopping injecting the gaseous ^{3}He, when there is still ^{3}He-^{4}He mixture in the reservoir.

One possible explanation is a superfluid ^{4}He film flow in the pumping tube which arises below 2.17 K. The film results from the siphoning and evaporates at the upper, warmer parts of the tube, which behaves as a thermal anchor between the coldest part and the warmer part of the cryostat. This process can cause both additional heat input and liquid vaporization.

The solution for this problem, without modifying the cryostat itself, is operating the Joule-Thomson stage with pure ^{3}He instead of ^{3}He-^{4}He mixture, which should result a lower base temperature of the cryostat.

Because the temperature depends only on the vapor pressure at liquid surface, it is necessary to calculate the pumping power as well as pressure distribution inside the pumping tube of the Joule-Thomson stage to estimate the base temperature using pure ^{3}He. As in the previous discussion, the ^{3}He flow should be $\sim 0.4\ \mu$mol per second. Thus one can obtain the Reynolds-number *Re* as:

$$Re = \frac{\rho v L}{\eta} \sim 500, \tag{3.2}$$

where ρ is the density of the ^{3}He [85], v the velocity, L dimension of the pumping tube and η the viscosity of ^{3}He [86, 87]. Because *Re* is less than 2000, the ^{3}He flow in the pumping tube is laminar. The pressure drop along the pumping tube is determined by [88]

$$\frac{\mathrm{d}p}{\mathrm{d}l} = \frac{128\eta\dot{V}}{\pi D^4}, \tag{3.3}$$

where $\dot{V}$ is the flux of ^{3}He and D the diameter of the pumping tube. Using the ideal gas approximation $p\dot{V} = \dot{n}RT$, one can get

$$\frac{\mathrm{d}p^2}{\mathrm{d}l} = \frac{256\eta}{\pi D^4}\dot{n}RT. \tag{3.4}$$

The viscosity η follows the mixture law [89],

$$\log\eta = N_{^3He}\log\eta_{^3He} + N_{^4He}\log\eta_{^4He}, \tag{3.5}$$

where N_i is the mole fraction of component i. The viscosity η is a function of temperature T, which depends on the position of the pumping tube l. The temperature distribution inside pumping tube $T(l)$ can be estimated by the temperature distribution in the liquid nitrogen bath-cryostat and liquid helium bath-cryostat (shown in Fig. 3.8(a)). Therefore all variables in the ordinary differential equation (3.4) have been explicitly written as a function of the position l, so that the integral can be easily numerically calculated from the position $l = l_0$, where the pump is installed and the end pressure $p(l_0)$ is known. The calculated pressure distribution is shown in Fig. 3.8(b).

One can see that there is almost no pressure drop (less than 1 Pascal) along the pumping tube. Therefore, the pressure as well as the temperature inside the Joule-Thomson stage is mainly limited by the power of the rotary pump which has a end-pressure of ~ 10 Pa with the gas load. Using the ^{3}He scale of temperature [82] the base temperature of the

Joule-Thomson cryostat operating with pure ^{3}He is estimated as 450 mK according to the pressure $p(l=0)$.

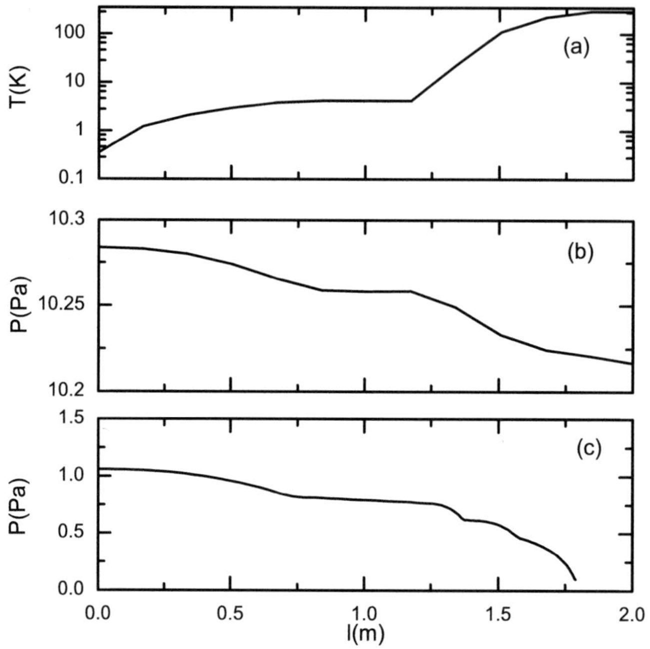

Figure 3.8: The temperature and pressure distributions inside the Joule-Thomson cryostat as a function of the position l in the pumping tube. $l=0$ is the place of ^{3}He pot. (a) the temperature distribution, (b) pressure distribution using a rotary pump. (c) pressure distribution using a turbo pump.

One can immediately see that, if a more powerful pump with lower end-pressure were installed to pump the ^{3}He, the base temperature could be even lower. A realistic idea is to use a turbo molecular pump, which has typical end pressure of ≤ 0.1 Pascal. The corresponding pressure distribution is shown in Fig. 3.8(c) and the estimated base-temperature is 350 mK.

Future development

Due to the obvious improvement of the performance as reducing the base temperature from 650 mK to 350 mK, the Joule-Thomson stage will be operated with pure ^{3}He instead of ^{3}He-^{4}He mixture and a turbo pump will be used for pumping the ^{3}He. A vibrational insulator

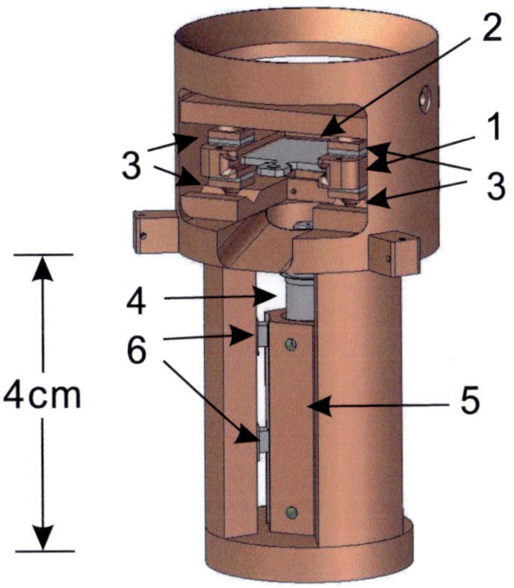

Figure 3.9: Scheme of STM-head, (1) sample rack, (2) sample and sample plate, (3) piezo for coarse motion in x-direction, (4) tip and scanner, (5) scanner-holder (Cu prism), (6) piezo for coarse motion in z-direction.

made of edge welded bellows has been constructed to support the turbo pump and reduce the vibrations. All these improvements will be tested in the near future.

3.2.2 STM head

The compact and stiff STM head was designed and modeled by finite element methods for its vibrational properties in order to reduce mechanical vibration. The microscope body was made of copper and copper beryllium. These have a high hardness and a high thermal conductivity. Figure 3.9 shows a schematic drawing of the STM head.

The samples are fixed by molybdenum springs on a 15x15 mm^2 square molybdenum sample plate, inserted into a sample rack made of copper beryllium at the upper part of the STM head. The STM scanner is a 6 mm piezo tube which is fixed in the scanner holder and has a maximum scan range of 2 μm x 2 μm and a z-range of 200 nm at voltages of $\pm$140 V and the base temperature. The STM tip is mounted in a tip holder which sits on top of the tube piezo.

The approach mechanism is based on the design by Pan *et al.* [47]. The scanner is fixed on a hexagonal Cu prism which is clamped by six ceramic isolators glued on top of six shear-piezos. The gliding planes of the prism were made from polished sapphire. Four of the shear-piezo pieces were directly glued on the main body of the STM, while the other two are pushed by a copper beryllium spring ensuring firm mechanical contact and sufficient force between the gliding planes of the slip stick motor. The travel in z-direction is 5 mm. Further, the Cu prism is thermally anchored to the body of the STM by a thin Au wire. The advantage of using a Cu prism is its high thermal conductivity thus ensuring a constant tip temperature during scanning and a fast cooling down of only a minute after extended coarse motion. Similarly, the sample rack can glide on polished sapphire plates using six legs, each made of a ruby sphere, shear-piezo, and a ceramic insulator. Therefore, coarse motion of the sample holder in x-direction is possible (range 5 mm).

3.2.3 Samples and STM tips

This experiment setup allowed *in situ* preparation of samples and STM tips.

The preparation chamber is equipped with an ion sputter-gun, allowing Ar^+ ion etching, several molecular beam evaporators, a quadrupole mass spectrometer, and a homebuilt electron bombardment heater for tips and samples with T_{max} >1500 K. For surface analysis, a combined low energy electron diffractiometer (LEED) and an Auger electron spectrometer were installed.

To avoid excessive warming during sample transfer, the warm sample is pre-cooled on a pre-cooling station attached to the radiation shield of the liquid nitrogen bath cryostat at a temperature of 77 K.

Using a molecular beam evaporator which is facing the STM head, *in situ* deposition of single atoms or molecules is possible without increasing the sample temperature above 12 K. Through subsequent cooling, the sample reaches the base-temperature within 10 minutes after the deposition procedure.

A modified sample plate is used to transfer STM tips. Only by force of the slip stick motors, the STM tip can be transferred into and out of the STM avoiding large mechanical forces on the scanner. This process is shown in Fig. 3.10. When the tip is on the modified sample plate, it can be transferred to the preparation chamber and can undergo all surface treatments that are available also for the samples.

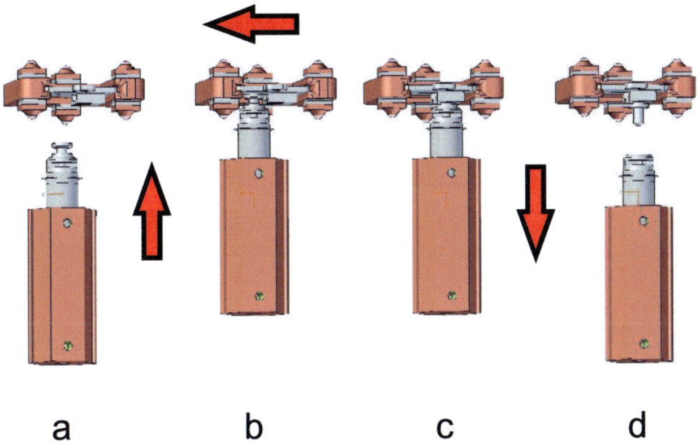

Figure 3.10: Process of tip transfer in four steps using a special sample plate and coarse motion in x and z directions.

3.2.4 Vacuum system

For surface science studies, an absolute cleanness of the investigated surfaces and total control of the composition of investigated materials are required. Thus all experiments must be performed in UHV.

To maintain the UHV, both chambers are continuously pumped by ion getter pumps. After baking the preparation chamber at 250 °C, a base-pressure of $<2\text{x}10^{-10}$ mbar was reached. A titanium sublimation pump is used to remove the reactive components in the residual gas. Since the cryostat is an efficient cryopump, a base-pressure of $<2\text{x}10^{-10}$ mbar was reached after baking the STM chamber at only 120 °C. The two chambers are separated by an UHV valve to avoid argon adsorption during ion sputtering. Further, the preparation chamber is connected to a load lock for quick sample and tip transfer into and out of vacuum without disturbing the UHV in the main chamber.

3.2.5 Molecular deposition

Unlike metallic atoms, organic molecules have lower sublimation temperatures (typically $80 \sim 300$ °C) and their thermal stabilities are much weaker. Therefore an evaporator with precise temperature control as well as heating power control is needed for the deposition of the organic molecules. In this work, a homebuilt molecular beam evaporator was designed specially for this purpose. Figure 3.11 shows a scheme of this setup. The molecules (b) in the Knudsen cell (c) are heated by a filament (d) and deposited to the surface of the substrate (a).

The temperature of the molecules is measured by a type-K thermocouple (e). By checking the signal from the thermocouple and adjusting the output of a current source (m), a PID controller (l) can stabilize the temperature of molecules with a accuracy of 0.1 °C.

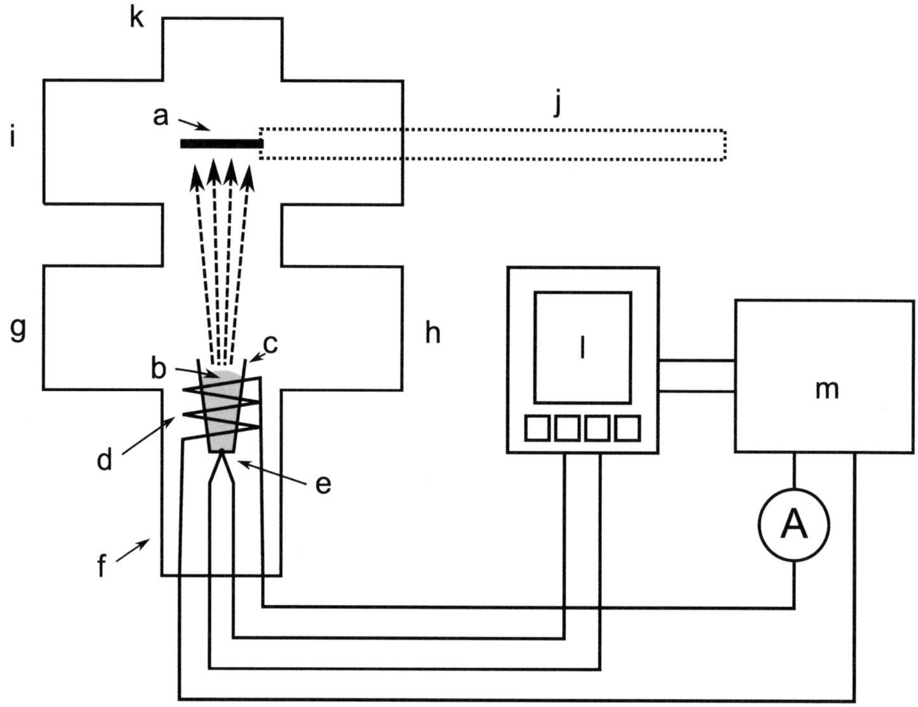

Figure 3.11: Scheme of molecular deposition. (a) sample, (b) molecules, (c) Knudesen cell, (d) filament, (e) thermocouple, (f) load-lock, (g) UHV ionization gauge, (h) turbo pump, (i) preparation chamber, (j) manipulator, (k) UHV viewport, (l) PID controller, (m) current source for the filament.

To avoid an interruption of UHV during the degasing of organic molecules, this evaporator was installed in the load-lock (f) pumped by a turbo pump (h). A UHV ionization gauge (g) is used for pressure control. Though the viewport (k) a change of the physical form of the molecules, like color, amount, can be observed. After deposition the sample (a) is transferred to STM by a manipulator (j) though the preparation chamber (i).

3.2.6 Electronics

The STM is controlled by commercial electronics [90] containing a PC-based real-time controller, a 22 bit multi-channel analog-to-digital/digital-to-analog converter, a high voltage power supplier, a piezo motor controller and a scanner controller. An analog lock-in amplifier is used when performing scanning tunneling spectroscopy (STS). The STM tip and sample are connected to a commercial preamplifier [91] mounted outside the UHV via shielded twisted-pair cables. To avoid high frequency noise, the STM bias voltage as well as the high voltage piezo signals are low-pass filtered with a cutoff frequency of 50 KHz.

The high voltage electric lines to the STM are NiMn wires with a low thermal conductivity, which are thermally anchored at 4.2 K and at the ^{3}He cryostat. The two coaxial lines for the tunneling current and the bias voltage are anchored in the same way. These cables are of low capacitance ($\approx$30 pF/m) and effectively damp high frequencies (-80 dB/m at 8 GHz and a 20 dB drop per octave) to hinder infrared photon conduction and to ensure low electron temperatures at the tunneling junction.

3.2.7 Vibration isolation

The tunnel current depends exponentially on the width of the vacuum barrier, which is typical only several Å during tunneling between the sample and STM tip. Thus, the STM head must be operated in an absolutely vibration-free environment to avoid the noise introduced by mechanically vibrations and the unexpected tip crashes which may damage or pollute the investigated nanostructures as well as the atomically sharp tip.

In order to achieve this requirement, the setup is supported by pneumatic isolators with laminar-flow damping to damp mechanical vibrations coming from the ground. Further, the STM head, which is mounted below the Joule-Thomson cryostat, is vibrationally isolated by three copper beryllium springs with resonance frequency of 2 Hz. All parts of the cryostat and the STM head were modeled with finite element methods to find their mechanical eigenmodes and frequencies. In an optimization process, the lowest eigenmodes were damped by mechanically suspension at the positions of vibration maxima or by thickening of critical parts. The pump for ^{3}He circulation was settled in a sound-proof case. These guaranties a z-stability at STM tip of $\approx$200 fm.

3.2.8 4 T superconducting magnet

For study of magnetism, it is advantageous to apply magnetic fields on a sample or STM tip [76].

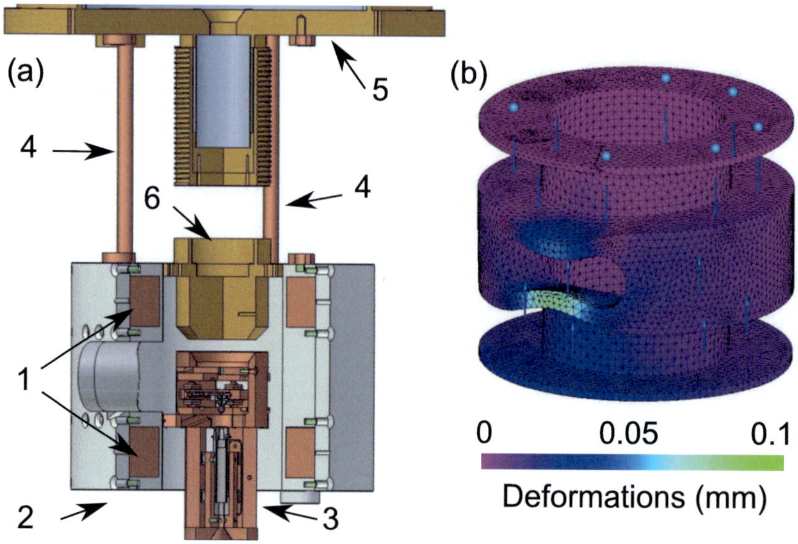

Figure 3.12: (a) Scheme of the superconducting split-coil magnet, (1) split-coil made of NbTi wire, (2) coil body, (3) STM head, (4) Cu posts as thermal anchoring between magnet and 4.2 K bath cryostat, which also support the coil body and carry the current for the coil back to ground, (5) bottom of 4.2 K bath cryostat, (6) ^{3}He pot of the Joule-Thomson cryostat. (b) Calculated deformations of the coil body when the coil generates a magnetic field of 4 T.

A superconducting split-coil magnet was mounted around STM head to apply an out-of-plane magnetic field (see Fig. 3.12(a)). The coil is made of one piece of 9 km superconducting NbTi wire with a diameter of 0.2 mm, which is able to carry current of 24 A under a 7 T magnetic field. The coil was embedded into UHV compatible Epotek H77 glue for stabilization. The magnetic field and the resulting mechanical deformations of the coil were modeled using finite element methods (see Fig. 3.12(b)). The field distribution is shown in Fig. 3.13. To minimize the heat input, high temperature superconducting wires are used to carry the current down from the liquid nitrogen bath cryostat to the superconducting coil. The coil body was made of Aluminum alloy (AlCuMg1) with high thermal conductivity and high mechanical stability. It is supported and thermally anchored to 4.2 K bath cryostat by three Cu posts. To prevent electrical shorts between the coil body and the superconducting wire, the coil body was anodized to generate a hard and isolating Al_2O_3 surface.

The superconducting split-coil is able to generate a maximal magnetic field of 4 T at the STM sample surface at currents of 20 A.

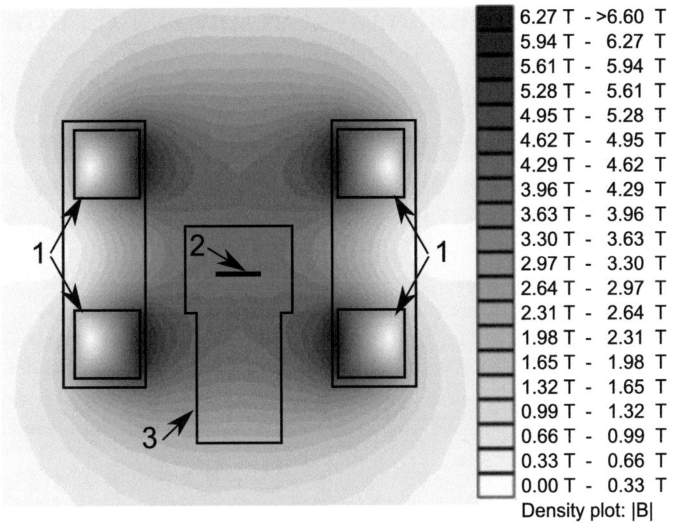

Figure 3.13: Calculated field distribution inside the split-coil magnet, (1) split coils, (2) sample, (3) STM head.

3.2.9 Temperature control

As described, the base temperature of the STM is 4.2 K without operation of JT-stage and 1 K (0.65 K) with JT-stage operating with ^{4}He (^{3}He-^{4}He mixture). However it is advantageous to allow to work at a variety of different temperatures (see Section 4.6.2). To achieve this, a low-temperature stable resistor of 2.2 K Ohm that can carry currents of up to 125 mA at room temperature was installed on the STM head as a heater. The heating power was controlled by a PID software module, which has been integrated into the STM electronics. With this, the temperature of STM can be controlled between the base temperature and $\sim$ 50 K with a high accuracy and stability ($\sim$ 2 mK). STM tests showed that there was no observable thermal drift when the temperature was maintained by the PID controller.

3.3 System test results

Several tests on the z-stability, noise level of the tunneling current, thermal drift, the superconducting magnet and the energy resolution of the STM, that have been achieved in the initial testing phase of the instrument, are presented in the following.

3.3.1 Z-stability and noise level of tunneling current

An in situ prepared Au(111) sample was used to calibrate the scanner piezo of the STM and to check the mechanical stability. The sample surface was bombarded with Ar^+ ions and annealed it at 730 K to achieve an atomically clean surface. High-quality STM images were obtained on this surface. The x-y calibration of the STM scanner was done by

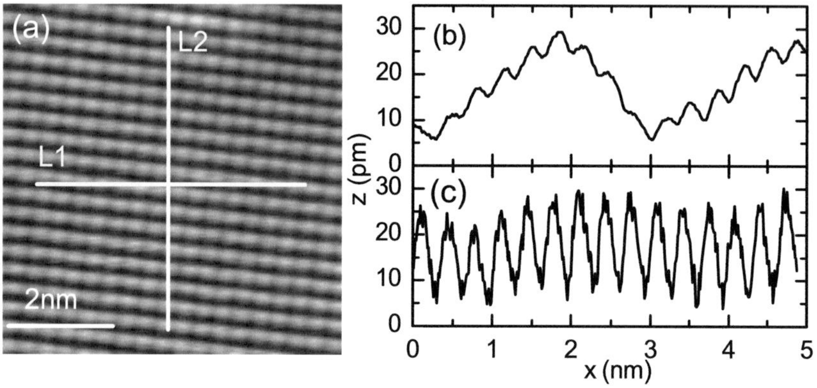

Figure 3.14: Atomic resolution of Cu(100) at 930 mK. (a) Topography of Cu(100), (b) line section along L1, (c) and L2.

measuring the reconstructions and the z-calibration was performed on atomic steps. Atomic resolution on a Cu(100) sample was achieved at 930 mK. (see Fig. 3.14). The line section of this topography shows an in-line z-noise level of well below 500 fm and a z-noise between-lines of below 5 pm. The main noise source in this case was the mechanical noise due to the bubbling of liquid nitrogen. It could be considerably reduced by pumping the nitrogen vapor to freeze the nitrogen or by filling the liquid nitrogen container additionally with fine copper wool. In the latter case, the high thermal conductivity of the Cu wool prevents the liquid from boiling from the bottom of the cryostat. Instead, the liquid evaporates only near the surface. Figure 3.15 shows separately the noise power spectra of the tunneling current at 1.1 K with liquid and frozen nitrogen in the cryostat. In case of frozen nitrogen the noise spectrum is below 6 fA/$\sqrt{Hz}$ at low frequencies which almost approaches the noise limit of the I-V converter of 4.5 fA/$\sqrt{Hz}$. A topography of the Au(111) surface (see Fig. 3.16) was taken at 930 mK without bubbling liquid nitrogen. Its line section demonstrates a reduced z-noise of about 200 fm.

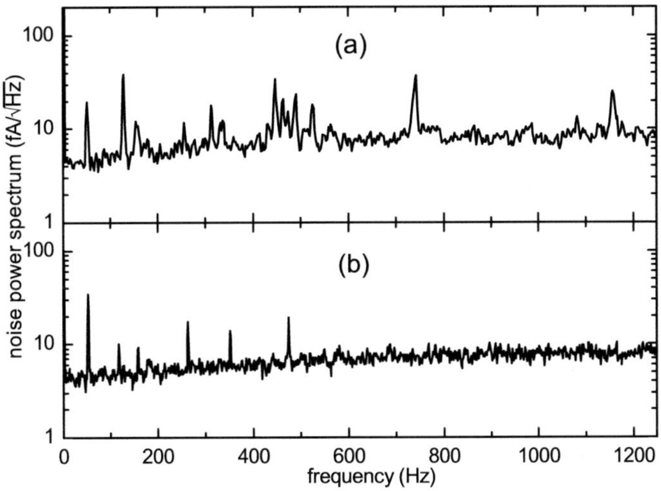

Figure 3.15: Noise power spectra of tunneling current at 1.1 K (a) with the liquid nitrogen bath fully filled and (b) with frozen nitrogen.

For some studies on insulators, semiconductors or molecules a tunneling current of several nA can damage the sample. Due to the low noise and high mechanical stability, this STM is able to investigate these fragile samples with low tunneling currents of several pA. A STM topography of the Au(111) surface with atomic resolution in constant-current mode with a tunneling current of only 1 pA and a sample bias of 1 mV is shown in Fig. 3.17

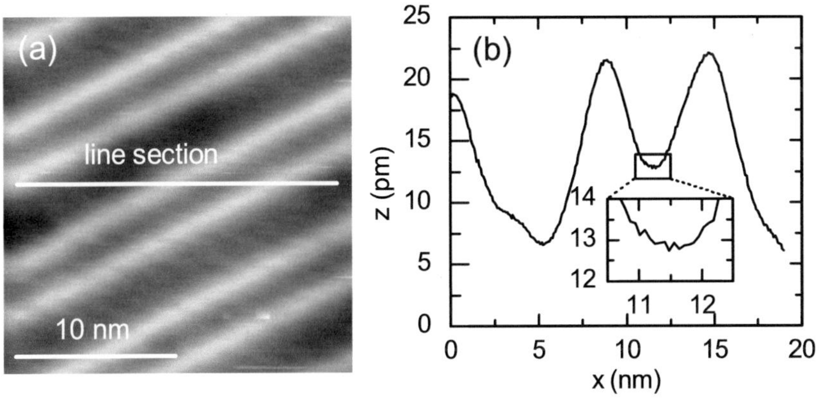

Figure 3.16: (a)Topography of the herringbone reconstruction on the Au(111) surface [92], (b) line section of (a) which demonstrates a z-noise of about 200 fm.

(a) . Similarly, scanning tunneling spectroscopy (STS) measurements with low currents of <4 pA were performed on organic molecules on Au(111). As shown in Fig. 3.17(b), the $I(U)$ curve was serially recorded by ramping the bias up (solid line) and down (dashed line). The two curves are almost identical and of low noise. To perform this measurement, the feed-back loop of STM controller was opened for three minutes. Therefore any drift or vibration between tip and sample during the STS measurement should be indicated by a separation of these two $I(U)$ curves. An evaluation of the differences between the two curves of currents in combination with distance current curves allows to extract a drift or vibration of <1 pm from the raw data, in good agreement with the vibration level observed during scanning.

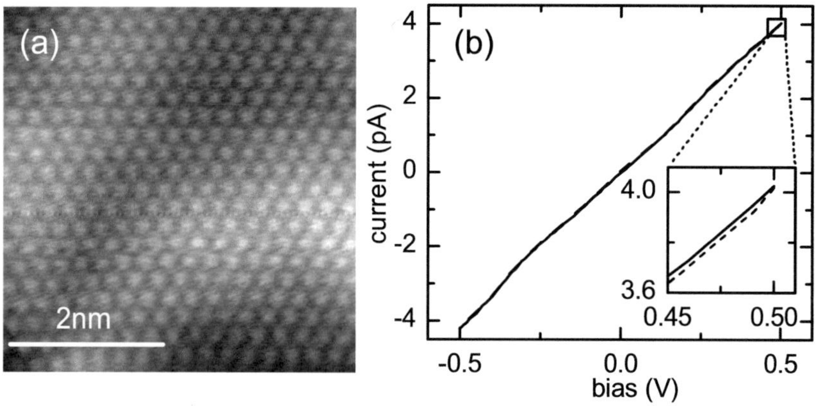

Figure 3.17: (a) Topography of Au(111) recorded at 1.1 K with a tunneling current of 1 pA and a sample bias of 1 mV. (b) STS measurement with a setpoint of 4 pA on a organic molecule adsorbed on Au(111) ramping the voltage up (soild line) and down (dashed line).

Finally the noise due to mechanical vibrations as well as voltage fluctuations on the piezo drivers was investigated. For this, we recoded a current noise spectrum with retracted STM tip and a noise spectrum with a tip in tunneling condition but with switched off feed back loop. Figure 3.18(a) shows the two current noise densities. The difference in these two originates from the current of the tunneling junction being sensitive to the gap widths, bias voltage instabilities and electronic shot noise. Using current distance curves (see Fig. 3.18(b)) this additional noise power can be converted in an equivalent vibration power shown in Fig. 3.18 (c). A high stability of only 2 fm/$\sqrt{Hz}$ is achieved.

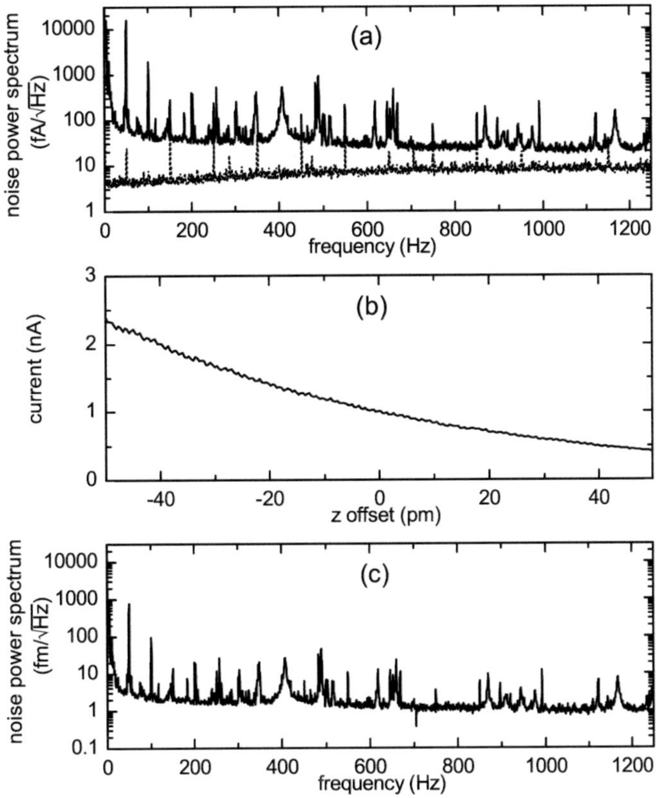

Figure 3.18: (a) Current noise power of retracted tip (dashed line) and in tunneling with turned off feedback loop (solid line). (b) Distance dependence of tunneling current I(d). (c) Extracted vibration noise power.

3.3.2 Lateral thermal drift

Due to the careful design of the cryostat and the STM body, the lateral thermal drift in STM measurements is negligible. Figure 3.19 shows four STM images of the same single Co atom on Au(111) which were taken serially at 0 minutes, 30 minutes, 100 minutes and 170 minutes after starting injection of ^{4}He into Joule-Thomson stage. The temperatures of STM head and positions of the Co atom were recorded as a function of time to determine the lateral thermal drift between sample and tip. The drift during cooling STM head from 4.2 K to 1.1 K was only 0.9 nm. After the temperature became stable the lateral drift in 2 hours was even smaller than the lateral resolution of STM image. This indicated that the lateral drift in the stationary situation is smaller than 20 pm per hour.

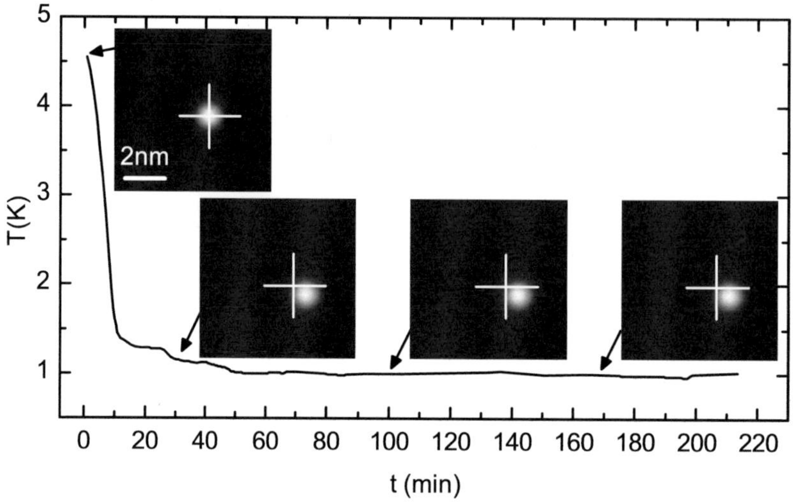

Figure 3.19: STM images of the same atom demonstrating no observable lateral drift (<20 pm per hour) after STM head reached 1.1 K. The drift was only 0.9 nm when STM was cooled down form 4.5 K to 1.1 K.

3.3.3 Flux vortices in Nb(110)

To test STM operation in a magnet field, a map of the quasiparticle density of states near the energy gap at $U = 1.2\ mV$ on a Nb(110) sample was performed at 1.1 K and a magnetic field of 0.21 T. The flux vortices, which are the characteristic of type II superconductors like niobium [93], were clearly resolved (see Fig. 3.20).

3.3.4 Energy resolution

To demonstrate the energy resolution STS measurements on in situ prepared Cu(100) were performed. A superconducting Nb tip that shows a superconducting phase transition at approximately 9.2 K was used. The Nb tip was chemically etched in a 1:1:1 mixture of HF, H_3PO_4, HNO_3 from Nb wire and was bombarded with Ar^+ ions and annealed at $\approx$1800 K to remove the oxide. Figure 3.21(a) shows a numerical differentiation of the current with respect to the sample bias measured at a temperature of 1.1 K. A superconducting energy gap is clearly resolved. Using conventional BCS theory [94] the gap in the spectrum was fitted. The temperature of the Fermi functions of tip and sample was also taken into account.

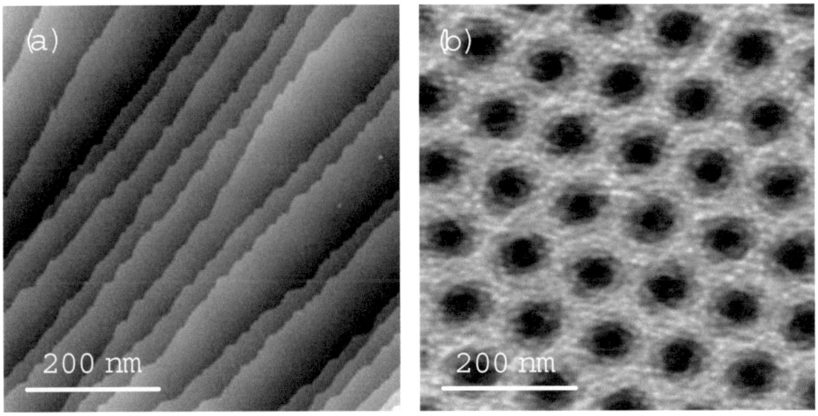

Figure 3.20: (a) Topography and (b) corresponding d^2I/dU^2 map of Nb(110) obtained at 1.1 K and a magnetic field of 0.21 T with a bias of $U = 1.2\ mV$. The flux vortices were clearly resolved in the map of the quasiparticle density.

The resulting fit is shown in Fig. 3.21 as the dashed line revealing a temperature of T_{fit}=1.15 K, which is close to the temperature read from the Cernox sensor on the STM head. Fitting the spectrum with focus on the height of the quasiparticle peaks revealed a lower temperature of 650 mK. The real temperature of the tip and sample is in between these two temperatures. This result indicates that the electronic temperature in spectroscopic measurements is also low and reveals an energy resolution that is given by the thermal limit of 3.4 kT_{fit}= 0.34 meV for dI/dU curves and corresponding 5.4 kT_{fit}= 0.54 meV for dI^2/d^2V curves.

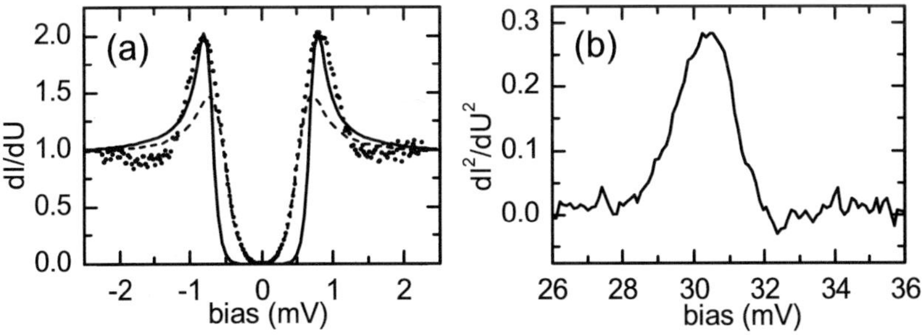

Figure 3.21: (a) STS of Nb tip on Cu(100), dotted line: experimental data, dashed line: fitting curve with a fit temperature of 1.15 K reproducing the energy gap; solid line: fitting curve with a fit temperature of 650 mK reproducing the quasi particle states. (b) Inelastic signal from the phonon excitation of CO on Au(111).

Finally, inelastic scanning tunneling spectroscopy (ISTS) was performed on a signal CO molecule on Au(111) using a tungsten tip. An inelastic peak of phonon excitation was clearly resolved in Fig. 3.21(b). The peak width at half maximum is only 2 mV. This is considerably sharper than previously reported for CO on Cu(111) [22] at similar temperatures.

4 Bipyrimidine-bridged homobinuclear complexes

As the bridge between two novel disciplines, spintronics and molecular electronics, single molecular magnets (SMMs) have attracted a lot of interest [54, 6]. The magnetic relaxation time of SMMs could be years at low temperatures, which has been observed in bulk crystals of SMMs [95, 96]. However, for applications like magnetic data storage, the SMMs have to be deposited onto a supportive substrate. Some higher order effects, like the Kondo effect, can appear in the system due to the hybridization of molecular states with the electrons of the substrate [8]. In this effect, the spin of SMMs is screened by the conduction electrons of the substrate [9] and the magnetic ground state of SMMs is lost. To avoid this effect, SMMs with two magnetic ions that couple antiferromagnetically were investigated. A series of binuclear metal-organic molecules of the form (M-(hexafluoroacetylacetonate)$_2$)$_2$-bipyrimidine- ((M-(hfacac)$_2$)$_2$-bpym), where M=Nickel, Manganese, Zinc, have been synthesized by Mario Ruben *et al.* at the INT, Karlsruher Institute of Technology (KIT). STM studies were performed on these complexes, the aim of which was to measure the exchange interaction between the two ions.

4.1 (M-(hfacac)$_2$)$_2$-bpym complexes

Figure 4.1 shows the structure of the (M-(hfacac)$_2$)$_2$-bpymcomplex. The molecule contains two metal ions M^{2+}, each of which is linked with two hexafluoroacetylacetonate (hfacac) ligands by M-O bonds. The aromatic ligand 2,2'-bipyrimidine (bpym) connects two of the M(hfacac)$_2$ components as bidentate chelate forming a complete molecule.

Three different types of molecules (M=Ni, Mn, Zn) of the (M-(hfacac)$_2$)$_2$-bpym complexes were studied in this work. In the following they will be called Ni_2, Mn_2, Zn_2. Although the molecules are similar, their crystal structures vary with the center ions. By performing crystal X-ray diffraction, the Mn(II) ions were found to be in a distorted trigonal prismatic coordination sphere and the Mn-Mn distance is 6.2 Å in Mn_2 dimer, while Ni(II) ions are in an octahedral environment [97, 98] and the Ni-Ni distance is shorter in Ni_2 being 5.6 Å (see Fig. 4.2).

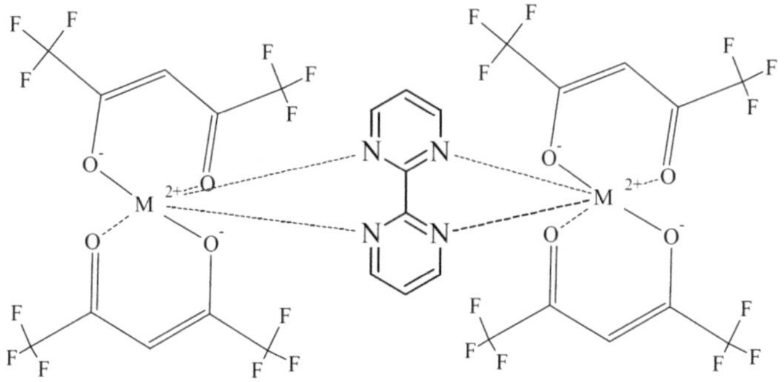

Figure 4.1: Structure of the (M-$(hfacac)_2)_2$-bpymcomplex. Each metal ion M^{2+} is linked with two hexafluoroacetylacetonate (hfacac) ligands by M-O bonds. By forming N-M bonds the bpym ligand coordinates as bidentate chelate with two of the M$(hfacac)_2$ components .

4.2 Magnetic properties

A weak antiferromagnetic exchange interaction between the two metal ions of the M-$(hfacac)_2)_2$-bpym complexes facilitated by the bpym ligand has been reported [97, 98]. The magnetic behavior of Mn_2 and Ni_2 as pure polycrystalline complexes has been measured between 2 K and 300 K by M. Ruben *et al.* [98].

Mn_2 has a room temperature magnetic moment of 8.7 $cm^3 \cdot K \cdot mol^{-1}$ corresponding to two uncoupled Mn ions with a spin value of $S = 5/2$ each, while Ni_2 has a room temperature magnetic moment of 2.12 $cm^3 \cdot K \cdot mol^{-1}$ and two uncoupled Ni ions with $S = 1$. Mn_2 and Ni_2 both show a paramagnetic behavior between 300 K and 75 K and a weak antiferromagnetic behavior below 75 K. The exchange interaction between two ions through bpym ligand is reported as J_{ex}=-1.6 meV [99] and J_{ex}=-2.06 meV [97] for Ni_2, which is about 10 times larger than the exchange J_{ex}=-0.13 meV of Mn_2 [98].

Because the exchange interaction between the molecules is relatively small compared to the thermal energy (~0.4 meV at 4 K and ~0.1 meV at 1 K) and the energy of the tunneling electrons (~-2 to 2 eV), according to the discussion in Section 2.2.2, the dimers can be excited by spin-flip scattering with tunneling electrons. Due to the conservation of

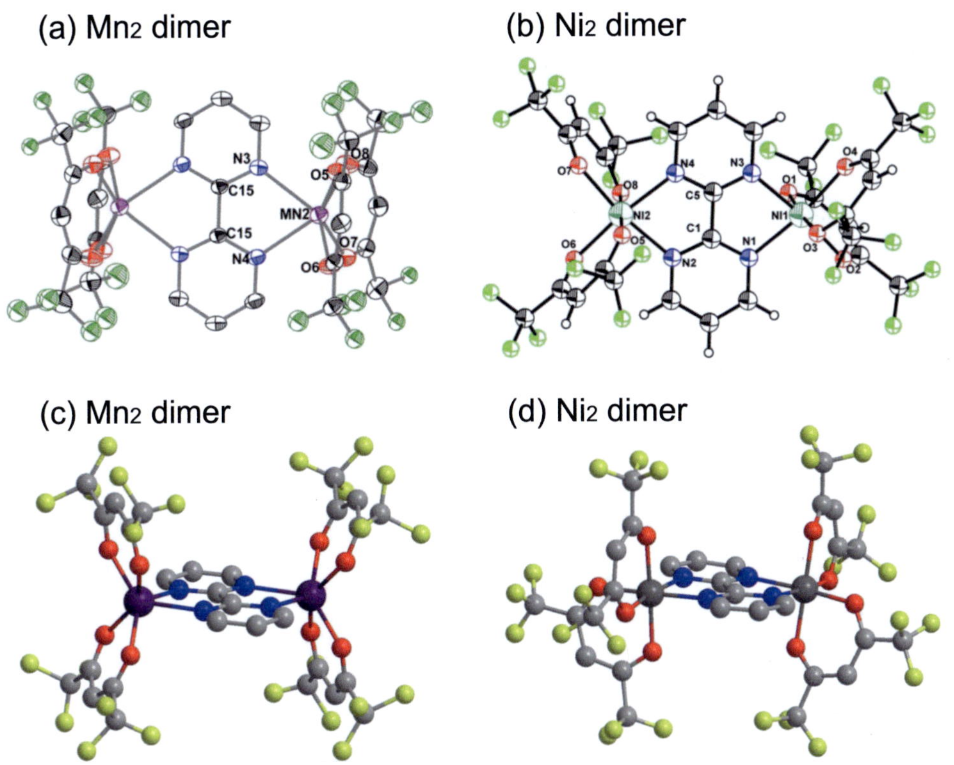

Figure 4.2: Structure of Mn_2 (a, c) and Ni_2 (b, d) obtained by single crystal X-ray diffraction . (a) (b) plotted by ORTEP with a "top down" view (ellipsoids drawn at %50 probability level), (c) (d) plotted by ChemBio3D with a "front" view. Hydrogen atoms are omitted for clarity.

angular momentum, the selection rule for the spin-flip excitation is $\Delta S = \pm 1$ and $\Delta m_s = \pm 1$. Considering the terms $s_i(s_i+1)$ in Equation (2.28) as constant, the Hamiltonian can be simplified as:

$$\mathscr{H} = -\frac{S(S+1)}{2} J_{ex}. \tag{4.1}$$

Thus the excitation energy from state $S = i$ to $S = i+1$ is

$$E_{ex,i \to i+1} = -\frac{(i+1)(i+2) - i(i+1)}{2} J_{ex} = -(i+1) J_{ex}, \tag{4.2}$$

which is shown in Fig. 4.3. As discussed in Section 2.1.4, these value can be obtained by performing ITS.

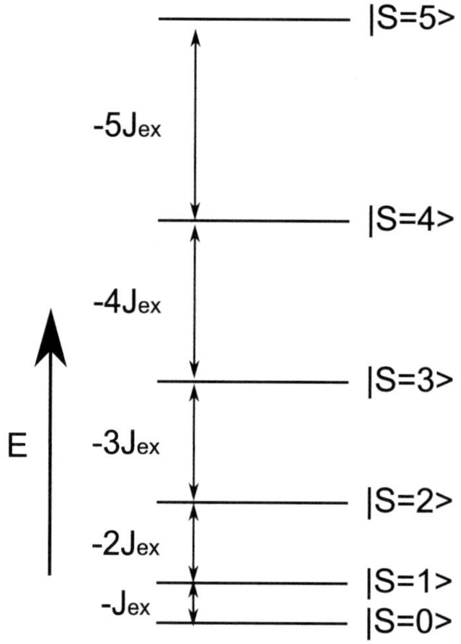

Figure 4.3: Scheme of the energy levels of Mn_2. Due to the conservation of the angular momentum the transition can only occur between two neighboring states. The energy levels of Ni_2 are similar except that the maximum spin is $S = 2$.

4.3 Degassing

To ensure a clean experimental conditions for the STM study, an atomically clean surface is needed and the components of the investigated molecules must be precisely controlled. After the synthesis, the (M-(hfacac)$_2$)$_2$-bpym complexes were purified by sublimation in rough vacuum at 140-180 °C, by which a purity over 98% could be achieved. However, during transport and storage, the molecules can still absorb some impurities from the atmosphere, for example H_2O. Therefore, the molecules must be purified again just before sublimating them to the substrate. The standard procedure to do this is heating the molecules in vacuum in a well-controlled way, so-called degassing.

The degassing of the molecules was done in the loadlock with the homebuilt molecular evaporator introduced in Section 3.2.5. The temperature of the molecules was increased carefully with a rate of 1-2 °C per minute by using the PID controller. During the warming up, the impurities were gradually removed from the molecules. This was checked by monitoring the pressure inside the loadlock. When the pressure reached an upper limit (typically 10^{-7} mbar), the heating was stopped and the molecules were cooled down. After the pressure getting better than 10^{-8} mbar, the molecules were warmed up again. By repeating this procedure, the impurities became less and less, and the pressure at the same temperate became lower and lower, therefore the final temperature of each cycle was higher and higher. When the pressure profile during each warming up became stable and the final temperature reached the sublimation temperature of the molecules, at which they sublime rapidly, the degassing procedure was stopped and the molecules were ready for deposition.

4.4 Mn_2 on Au(111)

The first system investigated in this work was Mn_2 on an Au(111) substrate. After sputtering with Ar^+ and annealing the sample at 400 °C, an atomically clean Au(111) surface was obtained (see Fig. 4.4 (a)). Before depositing Mn_2, the Au(111) sample was cooled down at the pre-cooling station in order to reduce the mobility of the molecules on the surface, so that each molecule could have a higher probability to stay on the surface as single object rather than forming islands. Then Mn_2 was sublimed at 135 °C and deposited on the Au(111) surface for 1 minute.

After transferring the Mn_2/Au(111) sample into the STM and cooling down to 4.2 K, STM measurements were performed. The topographies of the sample are shown in Fig. 4.4 (b)-(f). Unexpectedly, a self-oganisation of Mn_2 on the Au(111) surface was observed (b) (c), although the substrate was cold ($\sim$ 100 K) during deposition. When zooming in on self-organized structures (d) (e) (f), two layers of molecules could be recognized.

The aim of this work was to measure the exchange interaction between two ions in single (M-(hfacac)$_2$)$_2$-bpymmolecule. The magnetic properties of the molecules could vary in self-organized structures by the exchange interaction between the molecules, which is still unclear. After several attempts to obtain single molecules on the Au(111) substrate, it was proven that decreasing the deposited amount of the molecules by reducing the deposition temperature or shorten the deposition time, can only reduce coverage of the self-organized structure on the surface and cannot stop the molecules from forming islands or sticking at the step edges of the substrate. Furthermore, additional complexity due to the the herringbone reconstruction of Au(111) surface could be introduced during discussing the adsorption configuration of (M-(hfacac)$_2$)$_2$-bpymcomplexes on Au(111). Therefore, another substrate is needed onto which the molecule can be adsorbed as single objects.

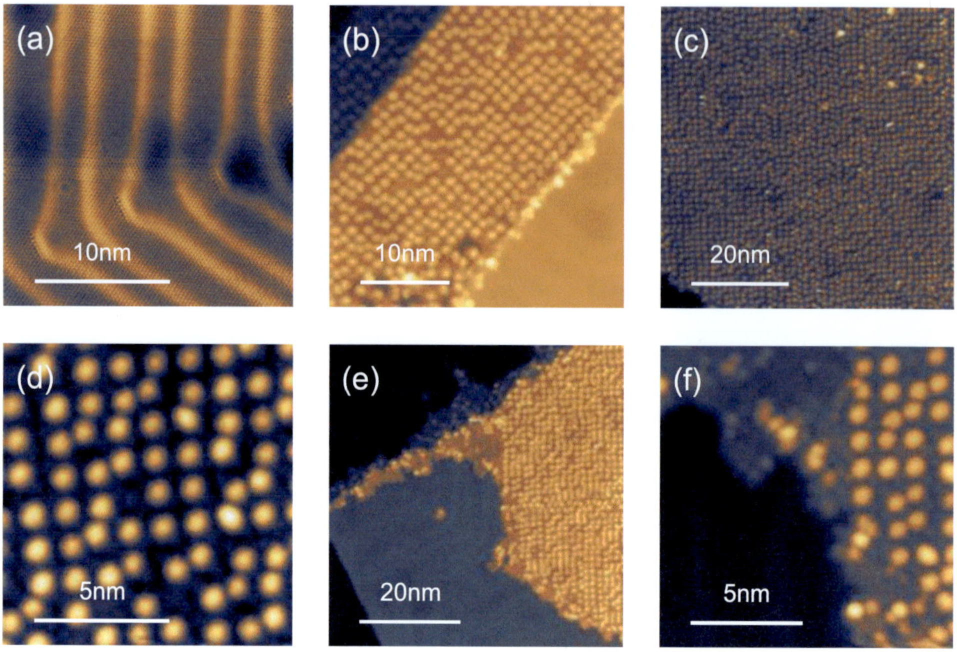

Figure 4.4: STM images of (a) Au(111) surface, (b)-(f) Mn_2/Au(111) taken at 4.2 K with a bias voltage between 0.75 and 1.75 V and a setpoint of the tunneling current of 5 pA.(a) Au(111) surface with atomic resolution, the herring bone reconstruction was clearly resolved, (b)-(f) self-organized Mn_2 on Au(111). From (d) and (f) one can recognize the well-ordered under lying layer.

4.5 Mn_2 on Cu(100)

According to the discussion in the previous section and the bipyramidal symmetry (mmm) of Mn_2, a Cu(100) surface with four-fold symmetry was used to support the molecules instead of Au(111).

4.5.1 Deposition and adsorption

After degassing, Mn_2 was deposited at 130 °C onto the atomically clean Cu(100) surface for 3 seconds. Before the deposition, the Cu(100) sample had been cooled down to $\sim$ 100 K at the pre-cooling station. Figure 4.5(a) shows a STM image of the atomic resolved Cu(100) surface by which the directions of the crystal were easily obtained. After deposition of Mn_2 about 60% of the surface was covered by islands formed by the molecules (see Fig.

4.5(b)). However some single objects with regular shape were found between the molecule islands (see Fig. 4.5(c)). The closeup views on these single objects (see Figure 4.5(d), (e)) show a centrally symmetric elongated structure, the long axis of which follows the [010] and [001] directions of the surface. By comparing the topographic images with the model of Mn_2 obtained by X-ray diffraction (see Section 4.1), these single objects fit in well with the structure of Mn_2: the long axis of the object corresponds to the direction of the Mn(II)-bpym-Mn(II) bond and the two bright spots at the ends of the long axis should be the two hfacac ligands of the complexes.

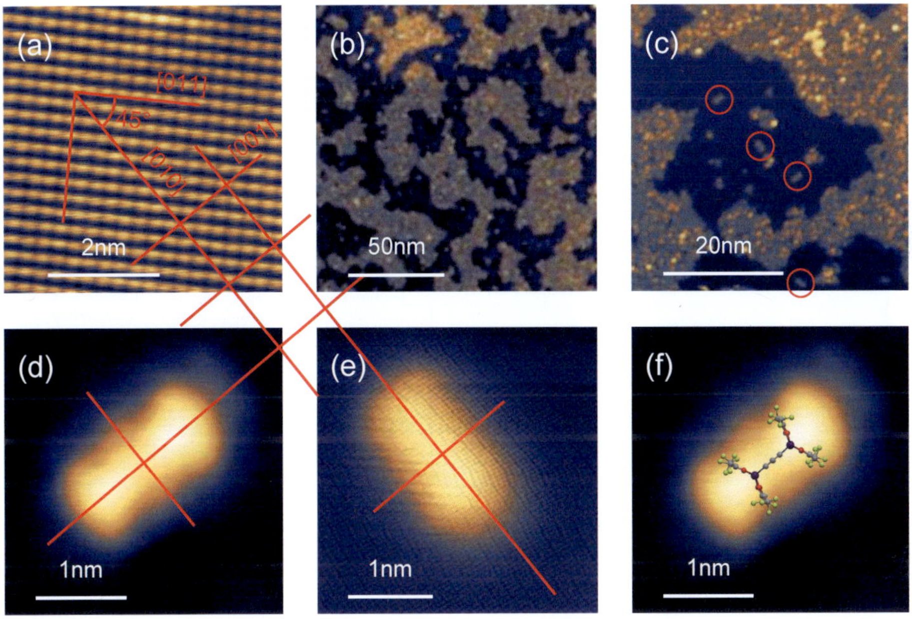

Figure 4.5: STM images of Mn_2/Cu(100) taken at 4 K. (a) Cu(100) surface with atomic resolution. (b) Cu(100) surface cover with deposited islands. (c) Single objects between the islands. (d), (e) Closeup of the single objects. Comparing with (a) one can see the long axis of them matching the [010] and [001] directions of the surface. (f) Matching the single object with the 3-d model of Mn_2.

4.5.2 STS of Mn_2/Cu(100)

After finding the single Mn_2 molecules on Cu(100), the next step was to investigate their possible excitations, especially magnetism-related phenomena (see Section 2.2.2, 4.2). As discussed in Section 4.2, the excitation energy of the antiferromagnetically coupled Mn(II) ions are $-J_{ex}$, $-2J_{ex}$, ..., $-SJ_{ex}$, where J_{ex} =-0.13 meV is the exchange interaction between the two ions and $S = 4$ is the maximum spin of this two-spin system. Therefore the expected excitations should be around 0.1~0.5 mV. The Mn_2/Cu(100) was further cooled down to 1.1 K by using the Joule-Thomson stage, so that a better energy resolution of 0.34 meV for dI/dU and 0.54 meV for dI^2/d^2U curves could be obtained, which is still insufficient to separate the peaks of the different excitations from $-J_{ex}$ to $-SJ_{ex}$. However, it is good enough to recognize if the peaks exist. STS (see Section 2.1.4) was performed on the single Mn_2 molecules. The results are shown in Fig. 4.6.

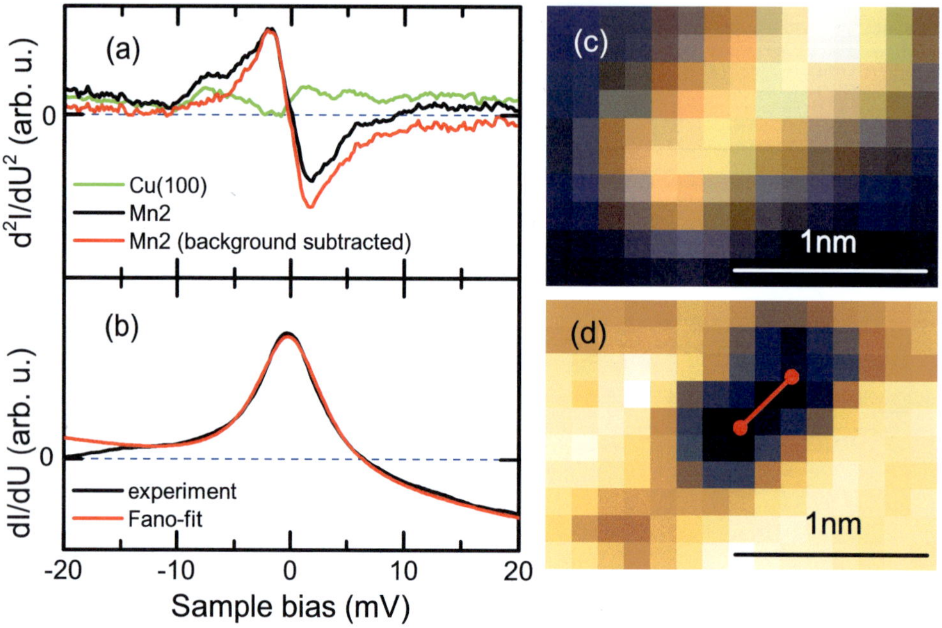

Figure 4.6: (a)The second derivative d^2I/dU^2 of STS of Cu(100) (green), Mn_2 (black) and background-subtracted STS of Mn_2 (red). (b) Differential conductance dI/dU (black) obtained by numeric integrating the d^2I/dU^2 (red line in (a)). By fitting with Fano resonance (red) a Kondo temperature of $T_K = 15$ K is obtained. (c) A topographic image of a single Mn_2 molecule. (d) A STS (d^2I/dU^2) map of the molecule in (c) at $U_{bias} = 1.8$ mV, where the dip of d^2I/dU^2 curve is, taken at 1 K with a modulation of $U_{mod} = 0.3$ mV. The two maxima of the STS correspond to the two Mn(II) ions.

By using a lock-in amplifier, the second derivative d^2I/dU^2 of the $I(U)$ curve of Mn_2 was obtained (black line in Fig. 4.6(a)) as well as the background signal d^2I/dU^2 of the Cu(100) substrate (green line in Fig. 4.6(a)). After subtracting the background, a clear peak-dip feature, which was unexpected, appeared near the Fermi level in the d^2I/dU^2 curve of Mn_2. Compared to a typical ITS signal, which shows a dip at negative bias and a peak at positive bias (see Fig. 2.3), this d^2I/dU^2 curve indicated an enhancement of the differential conductance near the Fermi level, which excludes a magnetic excitation.

Furthermore, this peak-tip feature in the d^2I/dU^2 curve had an asymmetrical shape, which is a hint for the appearance of the Kondo effect (see Section 2.2.3). The differential conductance curve dI/dU was calculated (black line in Fig. 4.6(b)) by numerically integrating the second derivative d^2I/dU^2. It shows a clear asymmetrical sharp peak at the Fermi level, which is the characteristic feature of a Kondo resonance. The appearance of the Kondo resonance is caused by the exchange interaction between the metal ions in the molecules and the conduction electrons in substrate [58]. More details about the exchange interaction could be obtained by determining the Kondo temperature T_K and the Fano parameter q which characterizes the coupling between the tip and the magnetic scattering centers.

According to Equations (2.30) and (2.31) the differential conductance curve dI/dU near the Fermi level can be fitted with the Fano resonance,

$$\frac{\mathrm{d}I}{\mathrm{d}U}(U) = A * \frac{(\frac{eU-\varepsilon_0}{\Gamma}+q)^2}{1+(\frac{eU-\varepsilon_0}{\Gamma})^2} + BU + C. \tag{4.3}$$

Considering the fact, that there is always some small constant offset from the lock-in amplifier and the dI/dU curve was obtained from the integral of the d^2I/dU^2 curve, an additional linear term $BU+C$ was used for the fit.

The result is drawn as red line in Fig. 4.6(b) with $\Gamma = 3.7 \pm 0.1$ meV and $\varepsilon_0 \sim 0$ (< 0.2 meV). The Fano parameter q has a valve of $\sim$100, which, however, has a huge uncertainty ($\sim \pm 500$) due to the additional degree of freedom introduced by the linear term BU.

Theoretically, the FWHM of the Kondo peak Γ is determined by the Kondo temperature T_K, the environment temperature T and the modulation U_{mod} from the lock-in amplifier. In this case the modulation $U_{mod} = 0.3$ mV is much smaller than $\Gamma = 3.7$ meV, so that the modulation can be neglected in the calculation. According to the (2.33), the Kondo temperature can be obtained by subtracting the influence of the environment temperature from Γ

$$T_K = \frac{1}{k_B}\sqrt{\frac{\Gamma^2/4-(\pi k_B T)^2}{2}} = 15.0 \pm 0.2\ \mathrm{K}. \tag{4.4}$$

4.5.3 STS map

The origin of the Kondo resonance is the scattering of conduction electrons and the isolated spins in the magnetic molecule. Therefore the spatial position where the Kondo resonance has the maximal amplitude should be related to the distribution of the magnetic scattering centers.

To resolve these maxima, a STS (d^2I/dU^2) map on a single Mn_2 molecule was recorded (see Fig. 4.6(c), (d)). During scanning over the whole molecule, the second derivative d^2I/dU^2 curve was measured on every points of the STM image. The position of maximal amplitude can be located by comparing the d^2I/dU^2 signal at a certain bias voltage. Figure 4.6(c) shows the STS map at $U = 1.8$ mV, i.e. the position of the dip on the d^2I/dU^2 spectrum, which leads to a maximal contrast of the d^2I/dU^2 signal.

Although the lateral resolution of the STS map was not optimized due to a limit of experimental time, one can still recognize that along the long axis there are two maxima near the center of the Mn_2 molecules with a distance of ~ 5 Å between them, which fits quite well to the two Mn(II) ions separated by a distance of 6.2 Å. This STS map gives another piece of evidence that the single objects on the surface are Mn_2 molecules and the enhancement of the differential conductance near the Fermi level is caused by the Kondo effect.

4.5.4 A brief summary

After deposition single Mn_2 molecules were found on the Cu(001) surface. However, in the STS experiments, instead of the magnetic excitation between the two antiferromagnetically coupled Mn(II) ions, the molecules show an unexpected Kondo resonance with a Kondo temperature of 15 K, which normally should not appear in an antiferromagnetic system. Before an in-depth discussion, this result needs to be compared with the measurements of Ni_2 and Zn_2.

4.6 Ni_2 on Cu(100)

As descried in previous sections, the Ni_2 and Mn_2 complexes are similar in their structures as well as in their chemical and physical (electric/magnetic) properties, while the exchange

J_{ex} between the two Ni(II) ions is one order of magnitude larger than in Mn_2. Therefore, a similar experimental procedure can be performed to investigate the magnetic properties of Ni_2 on Cu(100) in order to obtain single Ni_2 molecules on the surface. Similar experimental conditions should allow to compare the result from studies on different molecules in a meaningful manner.

Due to the larger exchange J_{ex}, the measurement of the magnetic excitation of the two antiferromagnetically coupled Ni(II) ions should be less restricted by the energy resolution of the STM, which is limited by the environmental temperature $T \sim 1$ K. However, it was unclear if a Kondo resonance could appear in this system just like in the case of Mn_2 on Cu(100).

4.6.1 Deposition and adsorption

After degassing, the Ni_2 molecules were deposited onto the atomically clean Cu(100) substrate with a sublimation temperature of ~ 100 °C. As shown in the previous section (see. Fig. 4.5), if the amount of the evaporated molecules is too high, most of the molecules form large islands and it will be difficult to find single molecules. Therefore, when depositing Ni_2 it was attempted to control the converge of the surface and the dimension of the islands formed by molecules by changing the temperature of the substrate during the deposition and tuning the sublimation temperature as well as the deposition time of the molecules. The topographic images of the surface are shown in Fig. 4.7. Surprisingly it reveals that also the adsorption configuration of Ni_2 could be influenced by the temperature of substrate during deposition.

Comparing the surfaces in Fig. 4.7(b) and (c), it can be seen that the coverage on the surface and the dimension of the islands could be reduced by lowering the evaporating temperature (105 °C vs. 100 °C) and the deposition time (20 seconds vs. 10 seconds). When the temperate of the Cu(100) substrate was kept as room temperature during the deposition of Ni_2, two types of single objects with regular shapes could be found among the islands on the surface (see Fig. 4.7(d), (e)), while only one type can be found when the substrate was pre-cooled in the STM before deposition (see Fig. 4.7(f)).

The two types of objects have the same four fold symmetry but a different shape and a different length of the long axis (1.3 nm vs. 1.9 nm). In the following, they will be mentioned respectively as 'equal-sign type' (the shorter objects marked by a green circle) and 'plus-sign type' (the longer objects marked by a red circle) for convenience. The orientation of the equal-sign type seems to be random on the surface, while the symmetry axes of the plus-sign type always follows the [011] and [01$\bar{1}$] direction of the substrate (see Fig. 4.7(a), (f)), like for the Mn_2. Considering this similarity and comparing the size of the single objects with the 3-d model, one can see that both of them should be the Ni_2 complex but with different adsorption configurations on the surface (see Fig. 4.8). While the plus-sign type 'lies' on the surface with both of its hfacac ligands, the equal-sign type 'stands' on the surface with one hfacac ligand and the other hfacac ligand is shown in its topographic image.

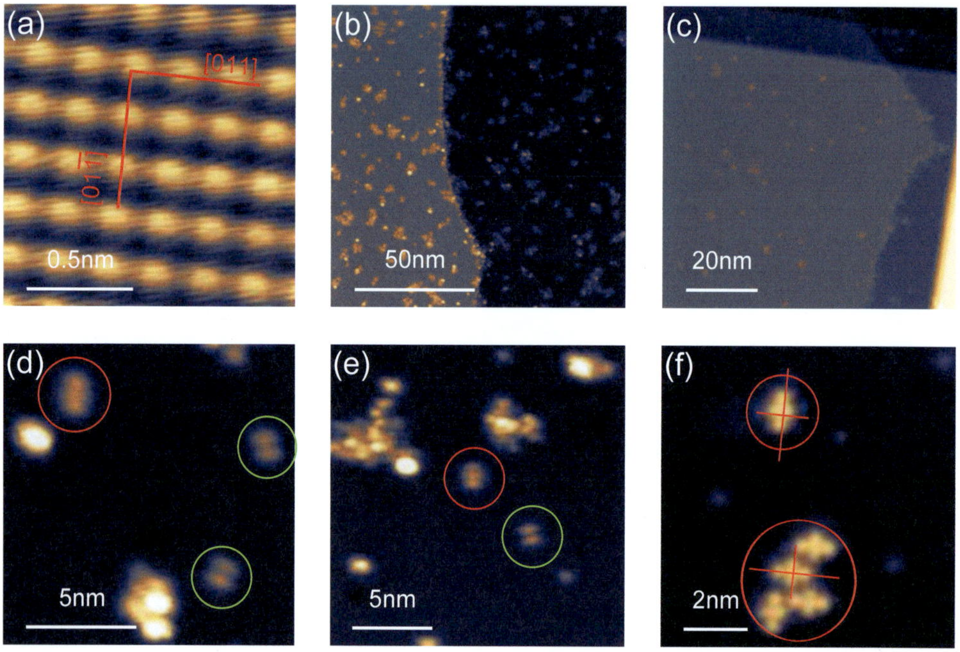

Figure 4.7: (a) Cu(100) surface with atomic resolution. (b) Ni_2/Cu(100) sample. Ni_2 was evaporated at 105 °C for 20 seconds. (c) Ni_2/Cu(100) sample. Ni_2 was evaporated at 100 °C for 10 seconds. (d), (e) Ni_2/Cu(100) sample. Ni_2 was evaporated when substrate was at room temperature. Two kinds of single objects could be recognized: (1) plus-sign type in red circle, (2) equal-sign type in green circle. (f) Ni_2/Cu(100) sample. Ni_2 was evaporated when substrate was pre-cooled in the STM. Only the plus-sign type can be found on the surface. The symmetry axes of the plus-sign type always follows the [011] and [01$\bar{1}$] direction of the substrate.

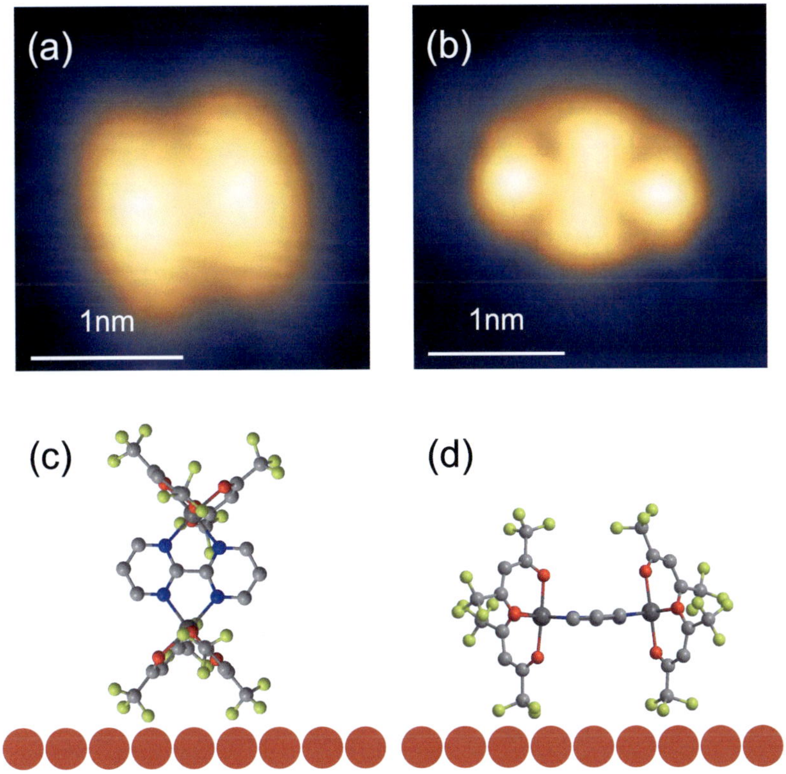

Figure 4.8: (a) STM image of equal-sign type, (b) STM image of plus-sign type, (c) scheme of the adsorption configuration of equal-sign type on the Cu(100) substrate, (d) scheme of the adsorption configurations of plus-sign type on the Cu(100) substrate.

4.6.2 STS of Ni_2 on Cu(100)

After deposition STS measurement were performed on the single Ni_2 molecules. The adsorption configuration dependence as well as the temperature dependence of STS was investigated.

STS of equal-sign type and its site dependence

The STS of the equal-sign type is shown in Fig. 4.9. The second derivative d^2I/dU^2 (a) and the first derivative dI/dU (b) (c) of the tunneling current were both recorded by a lock-in amplifier. As discussed in the previous section, ITS with a dip-peak feature on

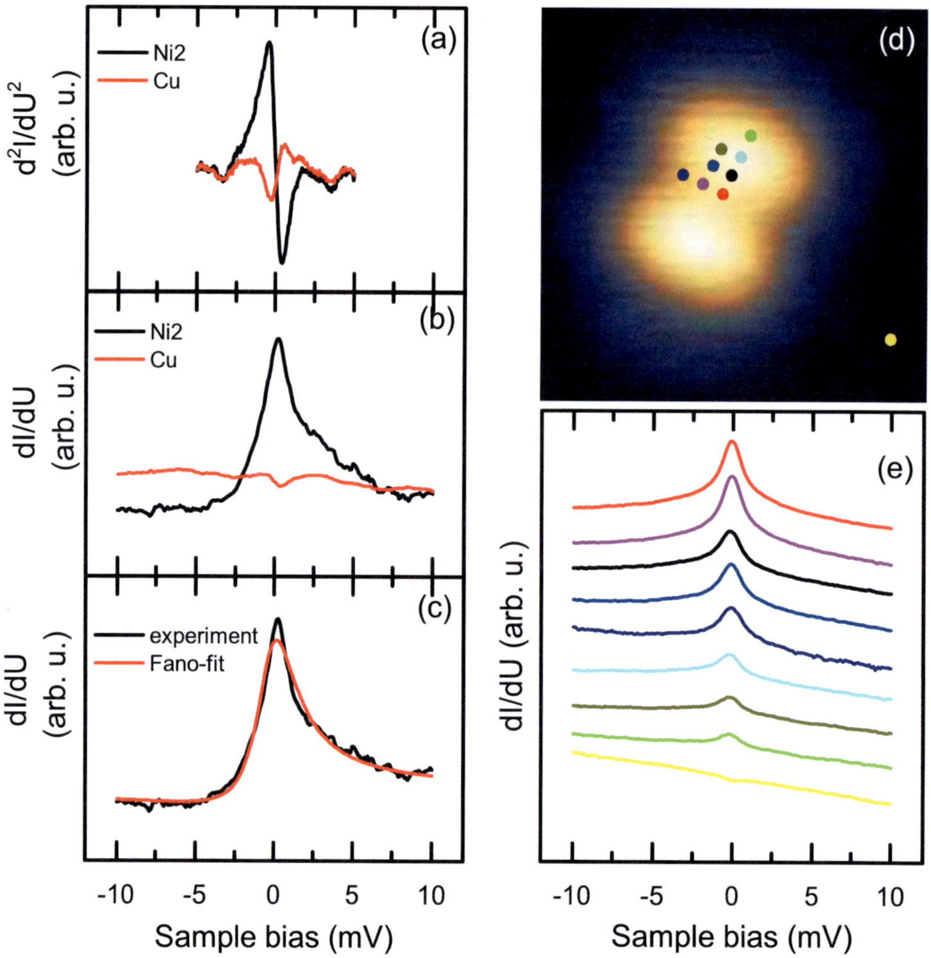

Figure 4.9: (a) Second derivative d^2I/dU^2 of the tunneling current taken at 1 K showing a peak-dip feature. U_{mod}=0.3 mV. (b) A sharp characteristic asymmetrical peak appears in the first derivative curve dI/dU at the Fermi level. T=1 K, U_{mod}=0.2 mV. (c) Background-subtracted dI/dU curve fitted by a Fano resonance with T_K=5.8 K and $q = 3.8$. (d),(e) STS on different positions of the molecule. The magnetic centers should locate at the place with the maximal amplitude of the Kondo resonance.

the second derivative curve indicating magnetic exciton of the two antiferromagnetically coupled Ni(II) ions was expected. However, a peak-dip feature near the Fermi level on the d^2I/dU^2 curve and a sharp unsymmetrical peak at the the Fermi level was observed on the dI/dU curve instead, which indicates a Kondo resonance arising in this system, just like in the Mn_2/Cu(100) sample.

After subtracting the background of the bare copper substrate, the dI/dU curve was fitted with a Fano resonance (see Fig. 4.9 (c)). The fit shows a Kondo temperature T_K of 5.8 K and a large Fano parameter of 3.8, which indicates a nearly full Kondo resonance between the tunneling electrons and the localized spin in Ni_2 [100].

By performing STS, it was found that the equal-sign type has a higher mobility on the Cu(100) surface than Mn_2, so that it can easily be laterally manipulated along the surface, when the tip applies a relatively high tunneling current of serval nA, which is necessary to obtain a sufficient signal-to-noise ratio for STS. This property results in a high possibility for the molecule to drift away from its original place after performing the STS and makes it difficult to obtain a complete STS map of the molecule with reasonable lateral resolution. However the dI/dU curve could be recorded manually as a function of the position of the STM tip (see Fig. 4.9 (d),(e)). The measurements show that the Kondo resonance has a maximal amplitude at the center of the molecules. Compared to the dI/dU map of Mn_2/Cu(100) sample (see Fig. 4.6), this result gives a hint of the position of the Ni(II) ions in the equal-sign type configuration, which supports the supposition that equal-sign type 'stands' on the Cu(100) surface.

STS of plus-sign type and the temperature dependence

The differential conductance curve dI/dU of the plus-sign type taken at 4 K is shown in Fig. 4.10. A sharp asymmetrical peak was clearly resolved at the Fermi level, which indicates that a Kondo resonance arises also in this configuration. Compared to the STS of the equal-sign type, one can see that the width of the Kondo peak of the plus-sign type is larger, which corresponds a higher Kondo Temperature T_K. This could be caused by a stronger exchange interaction between the conducting electrons and the localized Ni(II) ions in this configuration.

Besides the main peak around the Fermi level, STS showed two symmetric peaks at $U \simeq \pm$ 30 mV, so that it was impossible to fit the whole spectrum by a single Kondo resonance (see Fig. 4.10 (a)). One explanation for the two peaks is inelastic excitations such as phonon excitations or magnetic excitations. However, an inelastic excitation should result two symmetric steps in dI/dU curve rather than two symmetric peaks (see Section 2.1.4). The other possible explanation is a triplet-singlet Kondo resonance, which is a non-equilibrium Kondo effect [101]. A similar spectrum, which contains both zero-bias Kondo resonance and a triplet - singlet Kondo resonance at the same time in a two-spins system, has been observed by Mugarzal *et al.* [102]. By using this explanation, the STS was fitted and the results are shown in Fig. 4.10 (b), which exhibited a Kondo temperature of 16 K for the zero-bias Kondo peak and 45 K for the triplet-singlet Kondo peak. The exchage of the triplet-singlet transition was estimated as 22 meV.

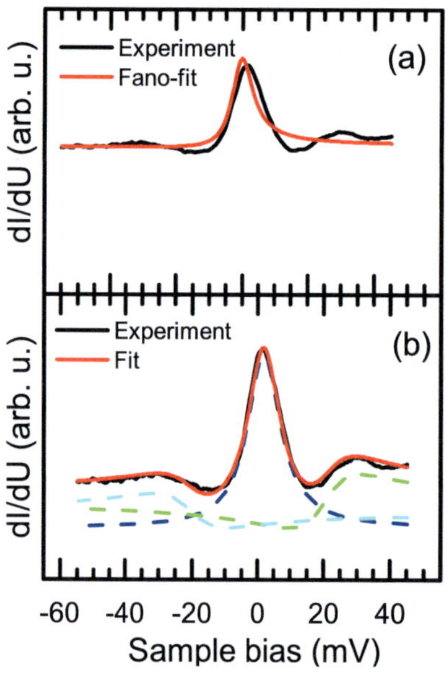

Figure 4.10: (a) Black line: The differential conductance curve dI/dU of the plus-sign type. Red line: fitting with a Fano resonance. (b) Black line: The differential conductance curve dI/dU of the plus-sign type. Red line: fitting with a superposition of zero-bias and triplet-singlet Kondo resonance. The blue dashed line: represents the peak at $U = 0$. The green and azure dashed line: represents the peak at $U \simeq \pm 30$ mV.

A dI/dU map of the plus-sign type was recorded at 1.1 K at $U = -10$ mV to investigate the position dependence of the Kondo resonance (see Fig. 4.12). Considering the enhancement of the conductance near the Fermi level due to the Kondo effect and the fact that the map was recored by using constant current mode, the position of the molecule with stronger Kondo effect should have a weaker dI/dU signal at $U = -10$ mV than other parts on the dI/dU map in this case. As shown in Fig. 4.12(b), two spots with weaker dI/dU signal separated by a distance of 4 Å are clearly resolved, which match with the two Ni(II) ions in the molecule quite well. This result gives a strong evidence that the plus-sign type are the Ni_2 molecules 'lying' on the Cu(100) surface.

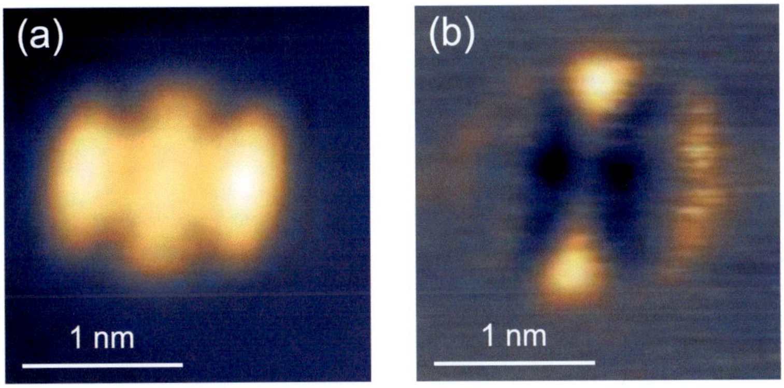

Figure 4.11: (a) A STM image of the plus-sign type, (b) corresponding dI/dU map, $U = -10$ mV, $I = 10$ nA, $U_{mod} = 2.5$ mV.

Because the Kondo temperature of the plus-sign type is much higher than the STM operation temperature, the temperature dependence of the Kondo resonance can be investigated by increasing the STM operation temperature without notable thermal broadening of the signal. The dI/dU curves at 0.65 K, 0.99 K, 2.0 K and 4.03 K was recorded respectively (see Fig. 4.12 (a)). Each of the curves was fitted as a superposition of zero-bias and triplet-singlet Kondo resonance and the energy width T_K and maximum intensity of the Kondo resonance G_K were obtained. One can see G_K decreases clearly with increasing of environmental temperature. As shown in Section 2.2.3, the temperature dependence of Γ and G_K follows the equations (2.33) and (2.34) respectively. Due to the large uncertainty of obtained parameters of the triplet-singlet Kondo resonance, only the Γ and G_K of the zero-bias peak were fitted with (2.33) and (2.34) and the results are shown in Fig. 4.12 (b), (c). G_K fits quite well the theoretical prediction while Γ shows a obvious aberration, which might be due to the large uncertainty of the fitting of Γ .

4.6.3 A brief summary

After deposition, two adsorption configurations of Ni_2 were found on the Cu(100) surface: a equal-sign type and a plus-sign type. Both of them show a Kondo resonance at the Fermi level but with different Kondo temperatures T_K and Fano parameters q. The plus-sign type with higher Kondo temperature shows a possible triplet-singlet Kondo resonance around a bias of $U = \pm 30$ mV. A site dependence STS study of the equal-sign type has proven that the molecule in this configuration 'stands' on the surface while the dI/dU map demonstrates that the plus-sign type 'lies' on the surface. The temperature dependence of the Kondo resonance of the plus-sign type shows that, the temperature dependent behavior of the maximal intensity of the zero-bias abnormity G_K fits well with the theoretical prediction, which is another evidence of the appearance Kondo resonance.

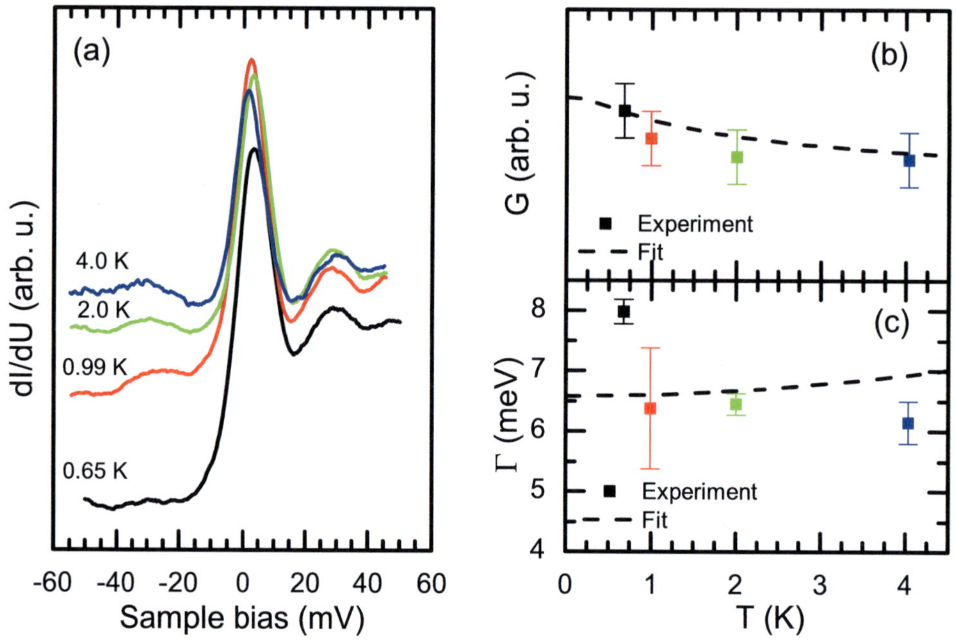

Figure 4.12: (a) The differential conductance curve dI/dU of the plus-sign type obtained at different temperatures. (b) Temperature dependence of the maximum intensity of the zero-bias Kondo resonance G_K. (c) Temperature dependence the energy width of the zero-bias Kondo resonance Γ.

4.7 Ni_2 on CuN/Cu(100)

In the previous sections it was shown that a Kondo resonance was observed in STS data of Mn_2 as well as of Ni_2 molecules on the Cu(100) substrate and the expected magnetic excitation of the two antiferromagnetically coupled ions was suppressed. Considering the fact that the appearance of the Kondo effect was caused by the exchange interaction between the Ni(II)/Mn(II) ions and the conduction electrons, one possible way to get rid of the Kondo effect is to reduce the coupling between Ni(II)/Mn(II) ions and the conduction electrons by depositing the Ni_2/Mn_2 complexes on a surface with an insulating layer instead of depositing them directly on the metallic surface. A usual way to obtain a surface with an insulating layer is to deposit Nitrogen atoms on a Cu(100) surface to synthesize ultrathin CuN films [103].

4.7.1 Growth of the CuN thin film and deposition of Ni_2

CuN films can be grown on Cu(100) by sputter deposition of atomic N followed by annealing at 330 °C. Depending on the deposition pressure, two kinds of CuN/Cu(100) surfaces can be obtained (see Fig. 4.13).

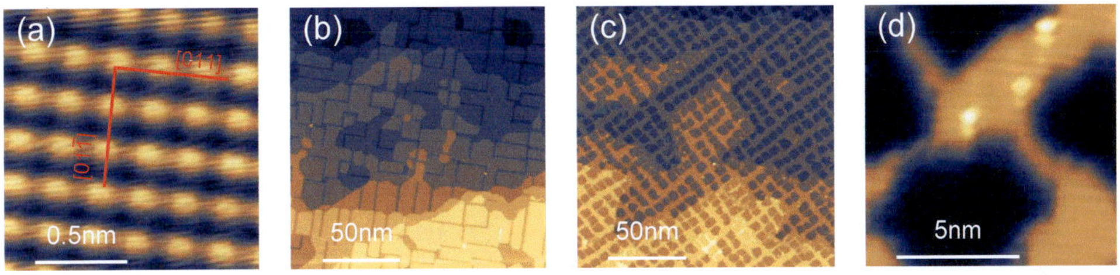

Figure 4.13: (a) STM image of bare Cu(100) surface with atomic resolution. (b) CuN/Cu(100) surface, nitrogen atoms were deposited with a pressure of $5x10^{-4}$ mbar and a deposition time of 10 min. (c) CuN/Cu(100) surface, nitrogen atoms were deposited with a pressure of $5x10^{-6}$ mbar and a deposition time of 30 min, (d) STM image of CuN islands with atomic resolution, U=1 V, $I_{setpoint}$=300 pA.

When the pressure of nitrogen was larger than $5x10^{-5}$ mbar during deposition, a CuN film with more than one CuN layer on Cu(100) can be grown with 100% coverage (CuN-rich type) (see Fig. 4.13(b)). Due to surface tensions the CuN layer on Cu(100), the substrate splits up into small pieces and the orientation of the trenchers follows the [011] and $[01\bar{1}]$ directions of the Cu(100) sample. When the pressure of nitrogen was lower ($<10^{-5}$ mbar), rectangular CuN islands with only one mono-layer thickness and a diameter of 5-10 nm grew on the surface (CuN-poor type) (see Fig. 4.13(c)). These islands appear as basins with a depression of 0.18 nm on the topographic images at low voltage [104]. Between the islands the bare copper can be seen and the edges of these rectangular islands follow the [010] and [001] directions.

After preparing the CuN/Cu(100) substrate, the Ni_2 molecules were deposited on to both kinds of the surfaces. However, the Ni_2 molecules could only be found on the CuN-poor-type surface (see Fig. 4.14). On the CuN-rich-type surface, there were no objects that could be recognized as Ni_2. On the CuN-poor-type surface, the Ni_2 molecules rather stick at the edges of the islands than staying in the central area of CuN islands or bare copper.

Compared with the STM images of Ni_2 on the bare Cu(100) surface in Fig. 4.7 and 4.8 one can see that all the Ni_2 molecules on the CuN-poor surface have the adsorption configuration of the equal-sign type. Depending on the individual adsorption geometries, most Ni_2 molecules lie mostly on the bare copper (see Fig. 4.14(b)) while some lie more on the CuN islands (see Fig. 4.14(c)). This implicates a different coupling between the molecules and the conduction electrons of the substrate which can result in different electronic and magnetic properties of the Ni_2 molecules on the surface.

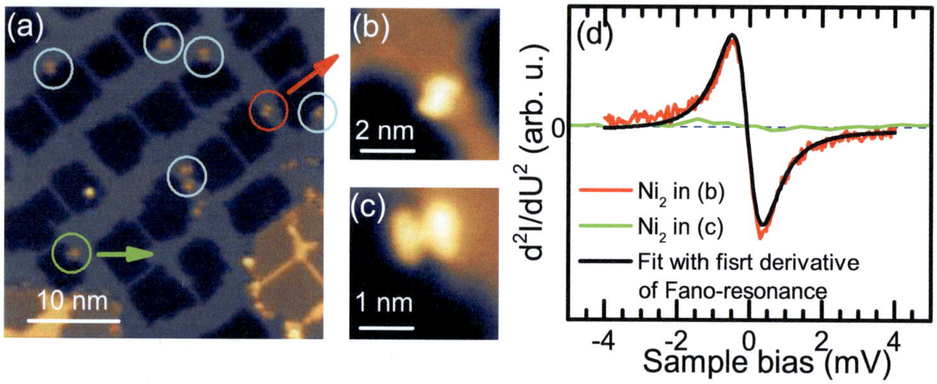

Figure 4.14: (a) STM image of Ni_2 molecules (marked by colored circles) on CuN-poor-type CuN/Cu(100) surface. (b) A Ni_2 molecule (marked by the red circle in (a)) lying mostly on bare copper. (c) A Ni_2 molecule (marked by the green circle in (a)) lying more on the CuN islands. (d) d^2I/dU^2 curves (Cu background-subtracted) performed on the Ni_2 molecule in (b) (red line) and in (c) (green line) taken at 1.1 K, $U_{mod} = 0.3$ V. One of the STS (red) was fitted with the first derivative of a Fano resonance (black line).

4.7.2 STS of Ni_2 on CuN/Cu(100)

After deposition STS was performed on the Ni_2 complex on CuN/Cu(100) substrate at 1.1 K. The d^2I/dU^2 curves are shown in Fig. 4.14(d). The Ni_2 molecules with different adsorption geometries obviously show a different behavior in STS. The d^2I/dU^2 curve of the Ni_2 molecule lying mostly on bare copper (Fig. 4.14(b)) showed a clear peak-dip feature at the Fermi level indicating the appearance of Kondo resonance just like in the case of the Ni_2/Cu(100) sample, while there was no remarkable signal on STS of the molecule lying more on the CuN islands (Fig. 4.14(c)). The d^2I/dU^2 curve, which showed a Kondo effect, was fitted with the first derivative of a Fano resonance. The fitting result demonstrates a lower Kondo temperature of only $T_K = 1.5$ K.

The origin of the different appearances of the Kondo resonance could be the different coupling between the Ni_2 molecules and the electrons in the substrate. When the Ni_2 molecule lies mostly on the bare copper, the spin of Ni(II) is screened by the electrons in the substrate, which increases the differential conductance at the Fermi level. When the Ni_2 molecule lies more on the CuN island, the interaction between the molecules and the substrate is weaker and the spin of the Ni(II) can hardly be screened by the electrons in the substrate (see Fig. 4.15(b)). The exchange coupling between the Ni(II) ions and the conduction electrons is cut off and the Kondo resonance is suppressed. This supposition of the tunneling process can be supported by the STM image in Fig. 4.14(c). The two ligands of the equal-sign type Ni_2 showed different color contrast and the height of the part of the molecule on the CuN island is smaller indicating a decease of the conductance, which means the coupling between the substrate and the ligands on the insulating layer was weaker than the coupling between the substrate and the ligands on the bare copper.

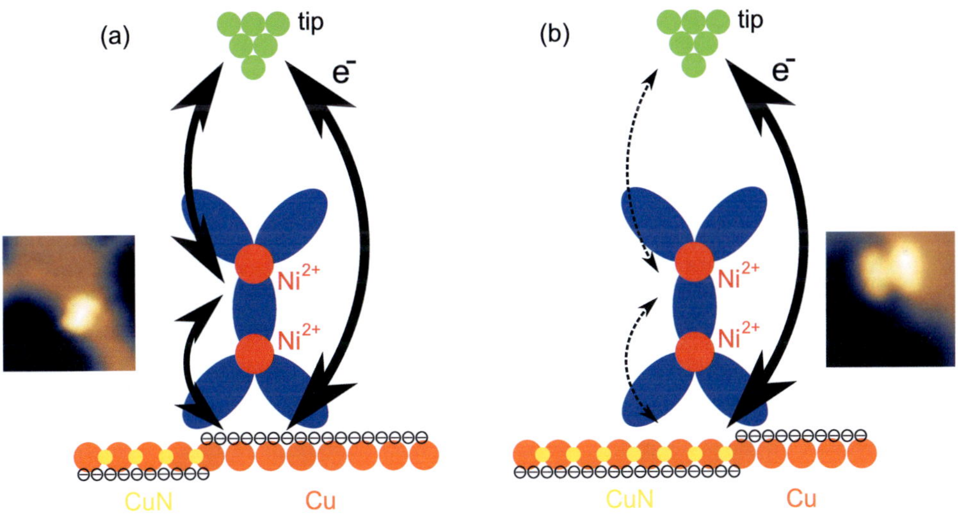

Figure 4.15: Scheme of how electrons tunnel into Ni_2/CuN/Cu(100) sample. (a) The Ni_2 molecule lies mostly on the bare copper. Due to the exchange between the ions and the conduction electrons in copper the Kondo resonance arises. (b) The Ni_2 molecule lies more on the CuN island with a weaker between the Ni(II) ions and the conduction electrons. The Kondo resonance is suppressed.

4.8 Zn_2 on Cu(100)

As a contrast experiment, Zn_2 complexes were also investigated in this work. Because the Zn(II) ion has a full 3*d* shell, the spin of the ion is zero. Therefore, there should be no magnetic excitations or Kondo resonance in the Zn_2/Cu(100) sample.

Zn_2 was deposited on the Cu(100) substrate at 85° for 30 seconds and three different adsorption configurations were found (see Fig. 4.16) on the surface. Like in the case of Ni_2, there are two adsorption configurations: (a) 'plus-sign type' and (b) 'equal-sign type' for the single Zn_2 molecules on the surface. Moreover, the Zn_2 formed also well-ordered islands on the Cu(100) surface (c) like Mn_2 on Au(111) (see Fig. 4.4(d), (f)) except that the Zn_2 islands were only one mono-layer thick.

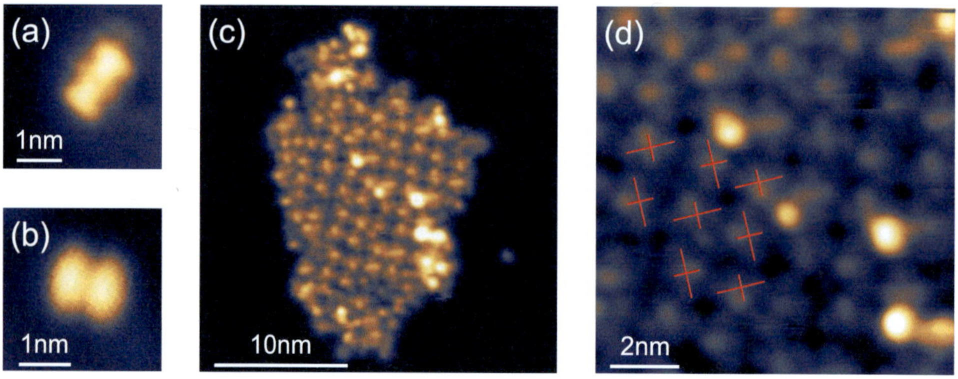

Figure 4.16: Topographic images of three different adsorption configurations of Zn_2 on Cu(100): (a) plus-sign type, (b) equal-sign type, (c), (d) one monolayer islands. In the islands, the Zn_2 molecules were ordered with a 1x2 pattern in which two molecules lay perpendicularly to each other in the surface plane.

STS was performed on the single Zn_2 molecules of both configurations. The results are shown in Fig. 4.17. As expected both of the dI/dU curves are flat near the Fermi level and there is no hint for magnetic excitations or Kondo resonance from +100 mV to -100 mV.

By comparing the STS of single Zn_2 molecules on Cu(100) with the STS of Mn_2/Ni_2, one can immediately see the obvious difference. The results of Zn_2/Cu(100) have proven that the signal found in the STS of Mn_2/Ni_2 complexes are not purely caused by the organic ligands but must be related to the properties of the Mn(II)/Ni(II) ions, the most remarkable one of which is the non-zero spin of them.

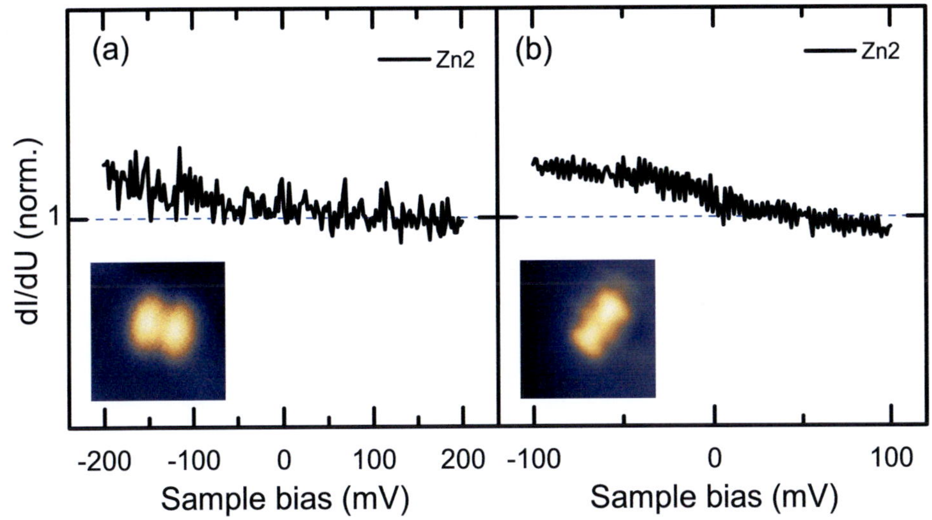

Figure 4.17: STS of the single Zn_2 molecules on Cu(100) surface. (a) STS of equal-sign type, (b) STS of plus-sign type. The dI/dU curves were normalized by comparing with the STS on bare Cu(100).

4.9 Conclusions and discussions

All three $(M\text{-}(hfacac)_2)_2$-bpym complexes (M=Ni, Mn, Zn) could be sublimed onto the metallic Cu(100) surface. Both Ni_2 and Zn_2 had two adsorption configurations as single molecules and only one was found for Mn_2. Ni_2 as well as Mn_2 showed a clear Kondo resonance signal near the Fermi level in STS while Zn_2 showed no feature. This difference is caused by the Mn(II) and Ni(II) ions which have non-zero spin. By fitting STS data with a Fano resonance, the Kondo temperature was estimated to be between 5 and 20 K depending on the type of the molecule and the different adsorption configurations. Ni_2 on CuN/Cu(100) showed that the Kondo effect could be suppressed by using an insulating layer which can separate the molecules and the metallic surface as well as reduce the coupling between localized ions and conducting electrons. However, the expected magnetic excitations were not found in any of these samples.

Generally speaking, the Kondo resonance can only arise in a system that contains two degenerate ground states. The $(M\text{-}(hfacac)_2)_2$-bpym (M=Mn(II), Ni(II)) complexes have two identical magnetic ions that are coupled antiferromagnetically by the exchange interaction via the bpym group with an integer total spin. So the ground state of the complexes should be a spin singlet state. Therefore a Kondo resonance should not appear in the Mn_2/Cu(100) and the Ni_2/Cu(100) sample, which is in conflict with the experiment.

One possible explanation is that after deposition on the substrate, the coupling between the Mn_2/Ni_2 molecules and the conduction electrons in substrate, which is characterized by the Kondo temperature T_K, is stronger than the antiferromagnetic coupling between the two Mn(II)/Ni(II) ions, which is characterized by the Néel temperature T_N. So the spin of single Mn(II)/Ni(II) ion is screened by the conduction electrons and Kondo resonance and the antiferromagnetic coupling between the two ions is cut off. Furthermore, because the crystal filed that the single ion can 'see' has almost no symmetry, all higher order terms in the Hamiltonian of the crystal filed can appear (see Equation (2.21)). Therefore, the possibility of a appearance of a Kondo effect is quit high.

5 Metal Acetylacetonates

Metal acetylacetonates (M(acac)$_n$, alternative name: Metal 2,4-pentane-dionate) are a class of coordination complexes, which have been intensively studied in last several decades, especially for the rare earth acetylacetonates [32, 33, 34, 35, 36, 37, 38, 39, 40], because they are the simplest known P-diketone chelates [37]. Their magnetic susceptibilities have been measured at low temperatures, and dysprosium acetylacetonate (Dy(acac)$_3$) showed a spin-lattice relaxation time of about a millisecond [34], which is a typical character of SMMs [6]. However, most studies were performed on powder specimens or in solutions. So far, studies on the growth and properties of metal acetylacetonates on surfaces are still rather sparse [41]. In this chapter low temperature STM studies on Dy(acac)$_3$ and Cr(acac)$_3$ are presented.

As shown in last chapter, magnetic excitations of magnetic molecules deposited on metallic substrate cannot be observed, when a Kondo effect arises. The precondition to get magnetic excitations is the stability of the molecular magnetic moment on the surface, which depends on the symmetry of the crystal field and total angular moment J [53]. The Dy(III) ion has a electron configuration of $4f_9$ with total angular momentum of 15/2 and a three-fold symmetry (C_3). According to the calculation (see Section 2.2.1), Dy(acac)$_3$ should have stable magnetic moment when it is deposited on the Cu(111) surface that has the same three-fold symmetry. Therefore, a magnetic excitation of Dy(acac)$_3$ on Cu(111) is predicted. As a contrast Cr(acac)$_3$ on Cu(111) was also investigated which could show a Kondo effect depending on its adsorption configuration.

5.1 Molecular structures

M(acac)$_n$ is a coordination complex consisted of one metal ion M^{n+} and n acetylacetonate ligands ($CH_3COCHCOCH_3^-$). Normally both oxygen atoms in the acac ligand bind to the metal ion to form a six-membered chelate ring (see Fig.5.1). Therefore, both Dy(acac)$_3$

and $Cr(acac)_3$ complexes have a coordination number of six and the center ions Dy(III) and Cr(III) are in the octahedral molecular geometry.

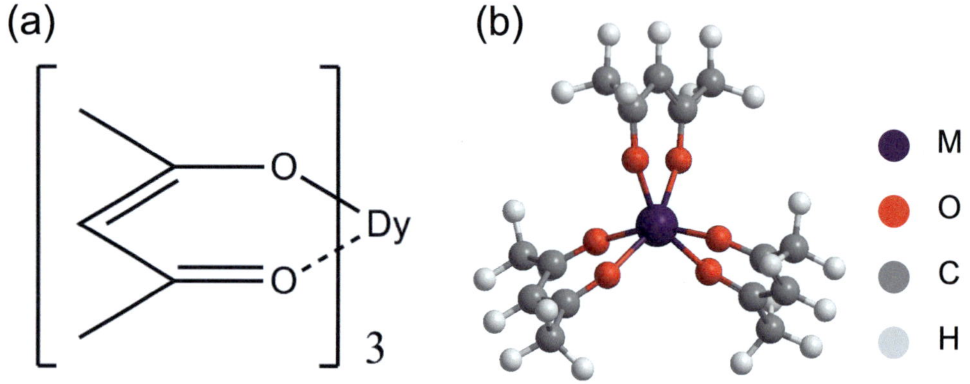

Figure 5.1: (a) Structural formula of $Dy(acac)_3$ (b) Ball-and-stick model of $M(acac)_3$ [105].

5.2 Experimental setup

A part of the following studies on $Dy(acac)_3$ were done in the group of Saw-W. Hla at Ohio University with a 4 Kelvin UHV STM and the studies on $Cr(acac)_3$ were performed in Karlsruhe with another 4 Kelvin UHV STM. Both of the two STM was cooled by a liquid helium bath cryostat with a basis temperature of 5-7 Kelvin. The base pressure in the chambers is $\sim 1\text{x}10^{-10}$ mbar after baking. All the samples were prepared *in situ* and the $Dy(acac)_3$/$Cr(acac)_3$ complexes were deposited on the substrate by using Knudsen cells. For the both setups, the typical noise level in z-direction at the STM tip is around 5 pm.

5.3 $Dy(acac)_3$ on Cu(111)

5.3.1 Degasing, deposition and adsorption

Compared to the Bipyrimidine-bridged homobinuclear complexes studied in the last chapter, the metal acetylacetonates complexes are relative easy to synthesize [105] and can be obtained commercially (grams instead of milligrams in the case of Ni_2/Mn_2) with a high

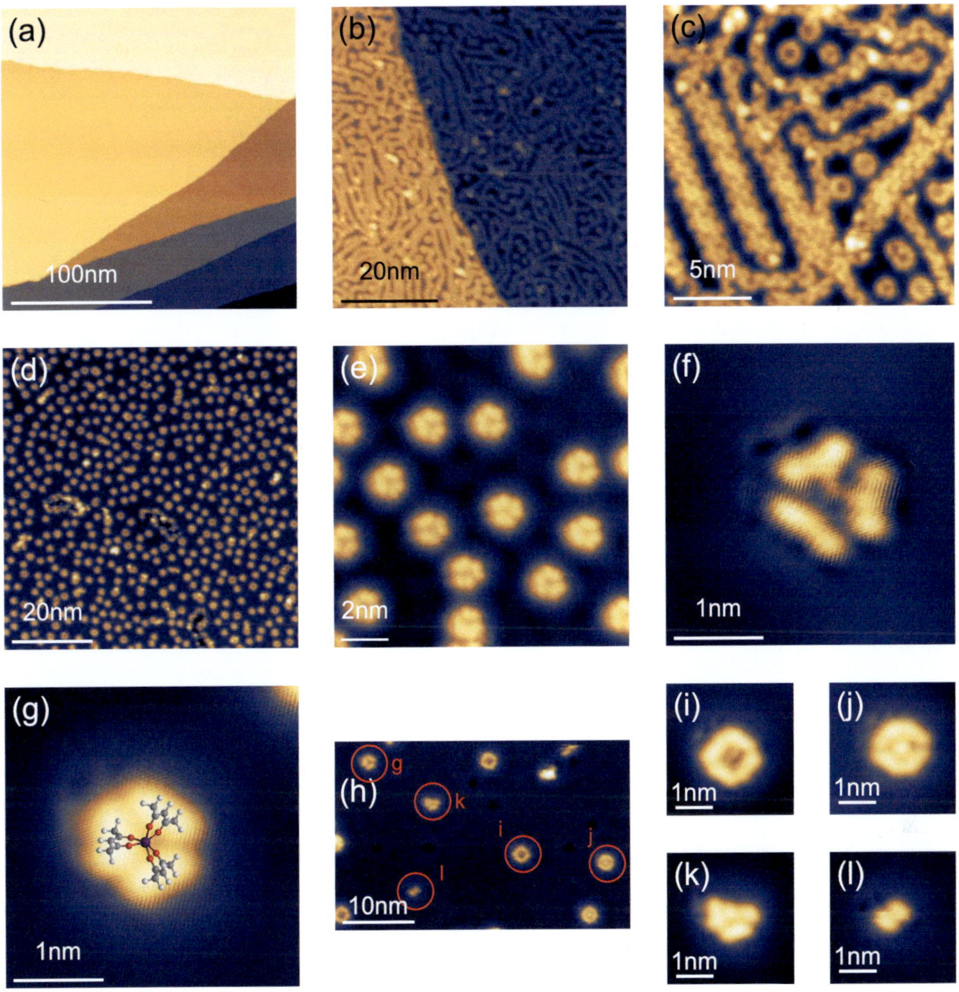

Figure 5.2: Topographic images of the samples. (a) Atomically clean Cu(111) surface. (b), (c) Cu(111) surface covered by molecules chains and few single molecules after depositing $Dy(acac)_3$ at 98°C for 1 second with a vapour pressure of $\sim 2\text{x}10^{-6}$ mbar. (d), (e) Cu(111) surface covered by single molecules with a coverage of $\sim$25%. (f) The dominant type of the molecules with a threefold symmetry. (g) Comparing the dominant type with model of $Dy(acac)_3$. (h)-(l) Other Types of single molecules that can be found on the surface.

purity over 99.9%. The melting point of Dy(acac)$_3$ is reported as 131-138 °C [33]. The long-term thermal stability and sublimation pressures of Dy(acac)$_3$ have been investigated by Siddiqi *et al.*, who showed that more than 80% of Dy(acac)$_3$ was not sublimed after heating to at 117 °C for 20 hours. It was also observed that a decomposition of the complex accompanied with the evaporation process takes place at lower temperatures. Due to the organic ligands, the rare-earth acetylacetonates are hygroscopic. However, the hydrate loses all H_2O when the pressure is lower than 1 mbar [33].

Considering all the above properties, Dy(acac)$_3$ was deposited on an atomically clean Cu(111) substrate at ~95°C for 1 second after degasing at 100 °C for serval hours in UHV. The Cu(111) substrate was precooled down to 20 K in STM before deposition. STM studies in this work shows that depending on the evaporation temperature and the vapor pressure in the chamber during deposition, Dy(acac)$_3$ has different adsorption configurations on the Cu(111) surface and, of course, different surface coverage (see Fig. 5.2).

When the vapor pressure in the chamber during deposition was $\sim 2\text{x}10^{-6}$ mbar, the coverage of Dy(acac)$_3$ after deposition was over 70%. Most molecules form zig-zag-chains. Only a few single molecules can be found among the chains (see Fig.5.2(b), (c)). When the vapor pressure was reduced to $\sim 3\text{x}10^{-7}$ mbar during deposition, only single molecules appeared one the surface (see Fig.5.2(d), (e)). The dominant type (over 80%) in these molecules show a threefold symmetry on the surface (see Fig.5.2(f)) and both their shape and diameter fit with the ball-and-stick model of the Dy(acac)$_3$ (see Fig.5.2(g)), which demonstrates that the adsorption configuration of Dy(acac)$_3$ is 'lying' on the Cu(111) surface.

Besides the molecules with threefold symmetry shown in Fig.5.2(f), (g), also some other types of objects have been found (see Fig.5.2(h)-(l)). Considering the Dy(acac)$_3$ is hygroscopic, some of the objects, such as the one in Fig.5.2(i), (j) showing more 'parts' than the molecules in (g), might be the hydrates Dy(acac)$_3$· nH_2O. The smaller objects, like the one in Fig.5.2(k), (l), might be the residue of the decomposition during evaporation process.

5.3.2 STS

After confirming that the individual Dy(acac)$_3$ molecule have been deposited on Cu(111) substrate, STS of single Dy(acac)$_3$ molecules was performed at 7 K as well as 0.8 K. The obtained dI/dU curves form -50 mV to 50 mV (a) as well as form -20 mV to 20 mV (b) are shown in Fig. 5.3. In the STS data two symmetric peaks at $U_{bias} = \pm 45$ mV are clearly

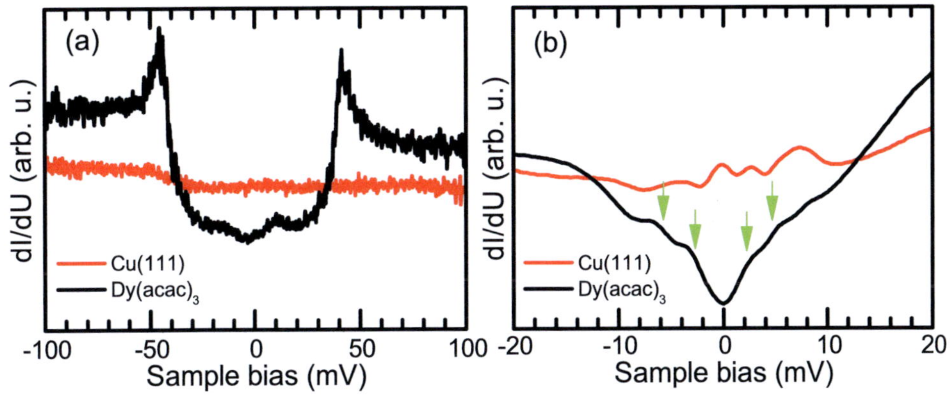

Figure 5.3: STS of a single Dy(acac)$_3$ molecule showing inelastic excitations at $U_{bias} = \pm 45$ mV, $\sim \pm 2.5$ mV and $\sim \pm 5$ mV, (a) from U_{bias}=-50 mV to U_{bias}=50 mV, taken at 7 K, (b) from U_{bias}=-20 mV to U_{bias}=20 mV, taken at 0.8 K.

resolved, which indicated an inelastic excitation feature. Furthermore, at the Fermi level a dip can be recognized which is clearly resolved by the STS taken at 0.8 Kelvin from -20 mV to 20 mV (b). This measurement demonstrates that two pair of steps located at $U_{bias} \simeq \pm$ 2.5 mV and $U_{bias} \simeq \pm$ 5 mV in the dI/dU curve exist, which correspond to two inelastic excitations with $E_{ex,1} \simeq 2.5$ meV and $E_{ex,2} \simeq 5$ meV.

In order to find out the origin of these excitations, the STS was taken as a function of positions (see Fig. 5.4) and a dI/dU map was recorded at $U_{bias} = 40$ mV (see Fig. 5.5). Both of the measurements showed that the largest amplitude of the STS appeared at the center of the Dy(acac)$_3$ molecules, where the Dy(III) ion is located. This demonstrates that the inelastic excitations are due to an interaction of tunneling electrons with the magnetic ions rather than the organic ligands.

usually, inelastic excitations below the energy scale of 10 meV are too low in energy for molecular vibrations, phonon excitations or plasmon excitations [106]. Considering that Dy(acac)$_3$ is magnetic molecule, a reasonable explanation for the inelastic excitations with $E_{ex,1} \simeq 2.5$ meV and $E_{ex,2} \simeq 5$ meV are magnetic excitations. As discussed in Section 2.2.1, the higher order terms in Hamiltonian of the crystal field can be neglected in lowest approximation, and only an uniaxial magnetic anisotropy along the z-axes of surface is left. In this case, the first excitation of a magnetic particle with an angular moment of J at low temperatures ($k_B T < E_{ex}$) is the excitation from the ground state $|J_z = J>$ to the first excited

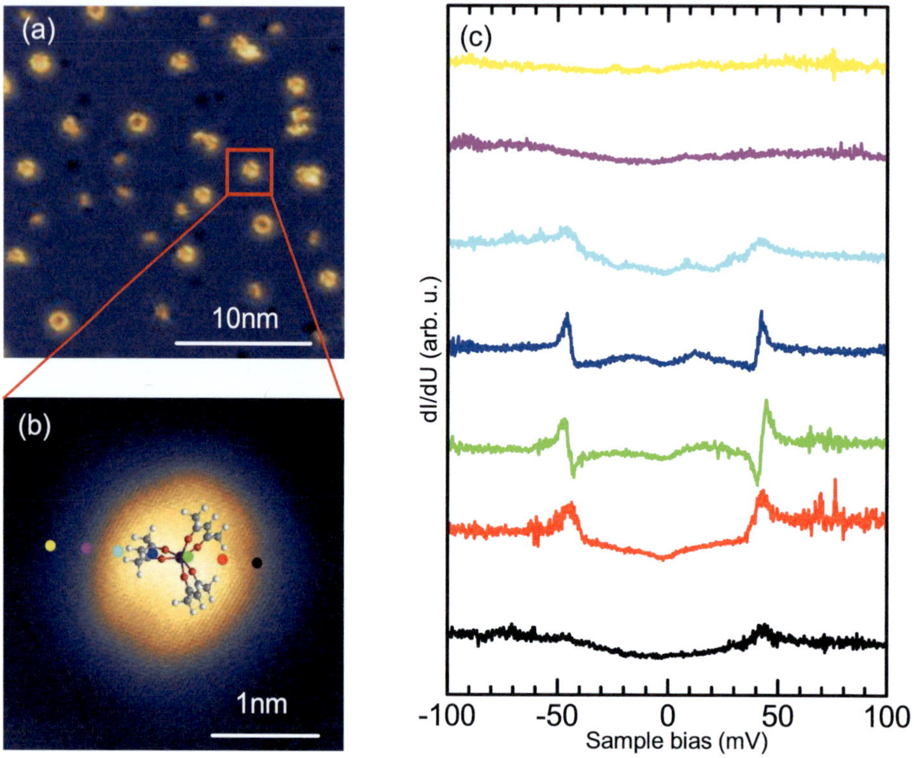

Figure 5.4: STS on the points along a line over a single Dy(acac)$_3$ molecule. The strongest signal of the inelastic excitations appeared near the center of the molecule.

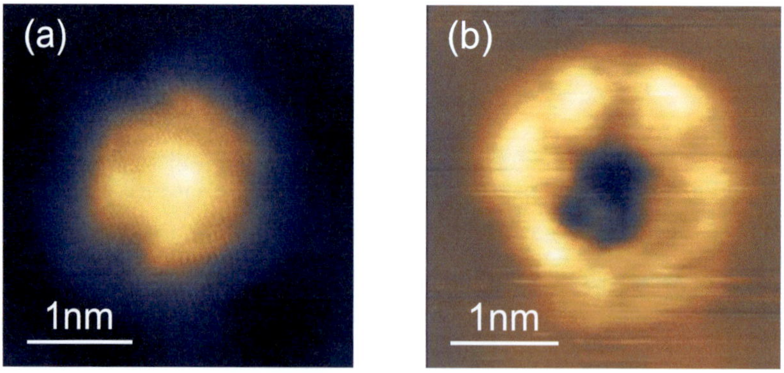

Figure 5.5: dI/dU map (b) of a single Dy(acac)$_3$ molecule (a) at U_{bias} =40 mV, where the dip of the STS is (I =0.7 nA T = 7 Kelvin). The strongest intensity of dI/dU was at the same position the Dy(III) ion.

state of $|J_z = J-1>$ the second excitation is from $|J_z = S>$ to $|J_z = J-2>$. According to (2.19) one obtains,

$$E_{ex,1} = D(J-1)^2 - DJ^2 = |D|(2J-1), \tag{5.1}$$

$$E_{ex,2} = D(J-2)^2 - DS^2 = |D|(2J+4). \tag{5.2}$$

In this case $E_{ex,1}$ is 2.5 meV and $E_{ex,2} \simeq 5$ meV and the Dy(III) ion has a electron configuration of $4f_9$, so the spin S should be 5/2, and the orbital angular momentum $L = 5$ because the orbital magnetic momentum of 4f electrons are usually not quenched. The total angular momentum J of ground state of Dy(III) ion follows Hund's rules and has a value of 15/2. So the uniaxial magnetic anisotropy $|D|$ and magnetic anisotropy energy (MAE) of Dy(acac)$_3$ on Cu(111) can be estimated as:

$$|D| = \frac{E_{ex,1} + E_{ex,2}}{(2J-1) + (2J+4)} \simeq 0.23\text{meV}, \tag{5.3}$$

$$\text{MAE} = |D|J^2 \simeq 12\text{meV}. \tag{5.4}$$

5.4 Cr(acac)$_3$ on Cu(111)

As a contrast experiment to Dy(acac)$_3$, a STM study of Cr(acac)$_3$ was performed. After degasing at 75 °C for 2 hours, the Cr(acac)$_3$ was deposited on clean Cu(111) substrate at 70 °C for 3 seconds. During deposition, the Cu(111) sample was at room temperature. The topographic images show that the self-organized chains of molecules form after deposition (see Fig. 5.6) while some single molecules can be found among the chains. Comparing the size and shape of the molecules with the 3d-model of Cr(acac)$_3$, one can see that the single objects as well as the self-organized chains are Cr(acac)$_3$ molecules that 'stand' on the Cu(111) surface with one acac ligand. In this case the Cr(III) ions 'feel' a crystal filed with twofold (or even less) symmetry. According to the discussions in Section 2.2.1, the possibility, that a Kondo effect appears in Cr(acac)$_3$/Cu(111) system, is quite high.

After deposition of Cr(acac)$_3$ on Cu(111) surface, STS were performed at 5 K. As expected an unsymmetrical peak appeared at the Fermi level which indicated a Kondo resonance (see

Fig. 5.7) and no remarkable difference could be found between the STS of single molecules and the chains. By fitting the dI/dU curve with the Fano resonance, a Kondo temperature of $T_K = 34 \pm 2$ K and a Fano parameter of $q = 0.014$ were obtained. Then the dI/dU maps were recorded on single molecules as well as on molecules chains (see Fig.5.8). The strongest signals of Kondo resonance always arose at the center of the molecules, where the Cr(III) ion is located, which shows the origin of the Kondo resonance is the central ions.

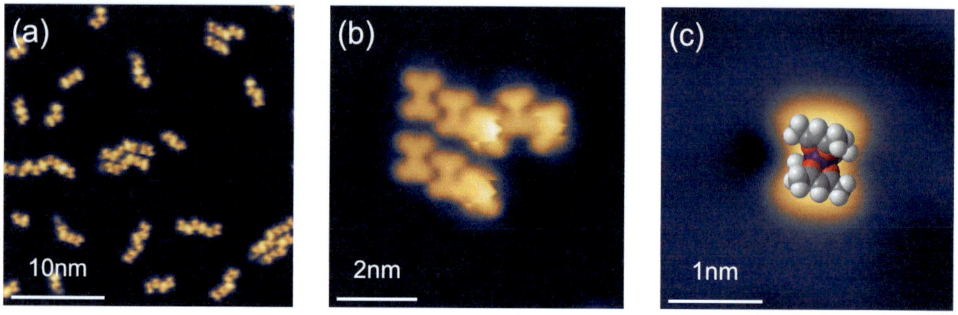

Figure 5.6: Topographic images of of $Cr(acac)_3$/Cu(111) sample. (a) Self-organized molecules chains. (b) Single molecules can be recognized in the chains. (c) A single molecules among the chains. By comparing with the 3-d model, the adsorption configurations of $Cr(acac)_3$ on Cu(111) was estimated.

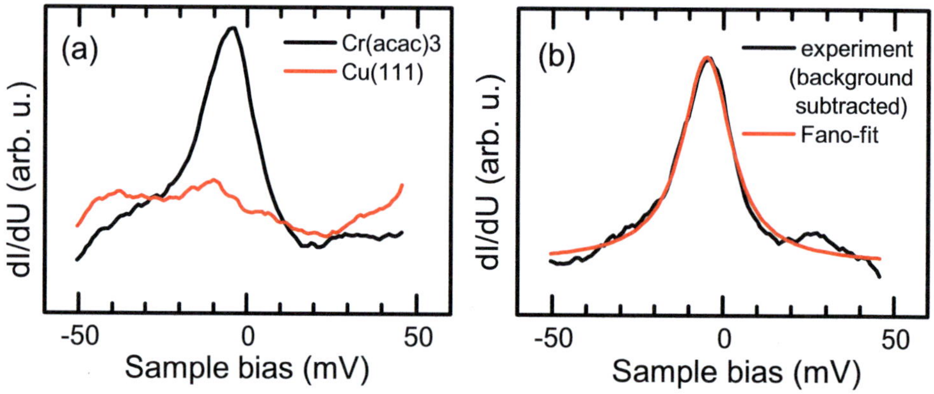

Figure 5.7: STS taken at 5 Kellin. (a) Black line: dI/dU curve on a single $Cr(acac)_3$ molecules, red line: on the bare copper substrate. (b) After subtracting the background signal of bare copper, the STS of $Cr(acac)_3$ molecules was fitted with a Fano resonance with a Kondo temperature $T_K = 34 \pm 2$ Kelvin and a Fano parameter $q = 0.014$.

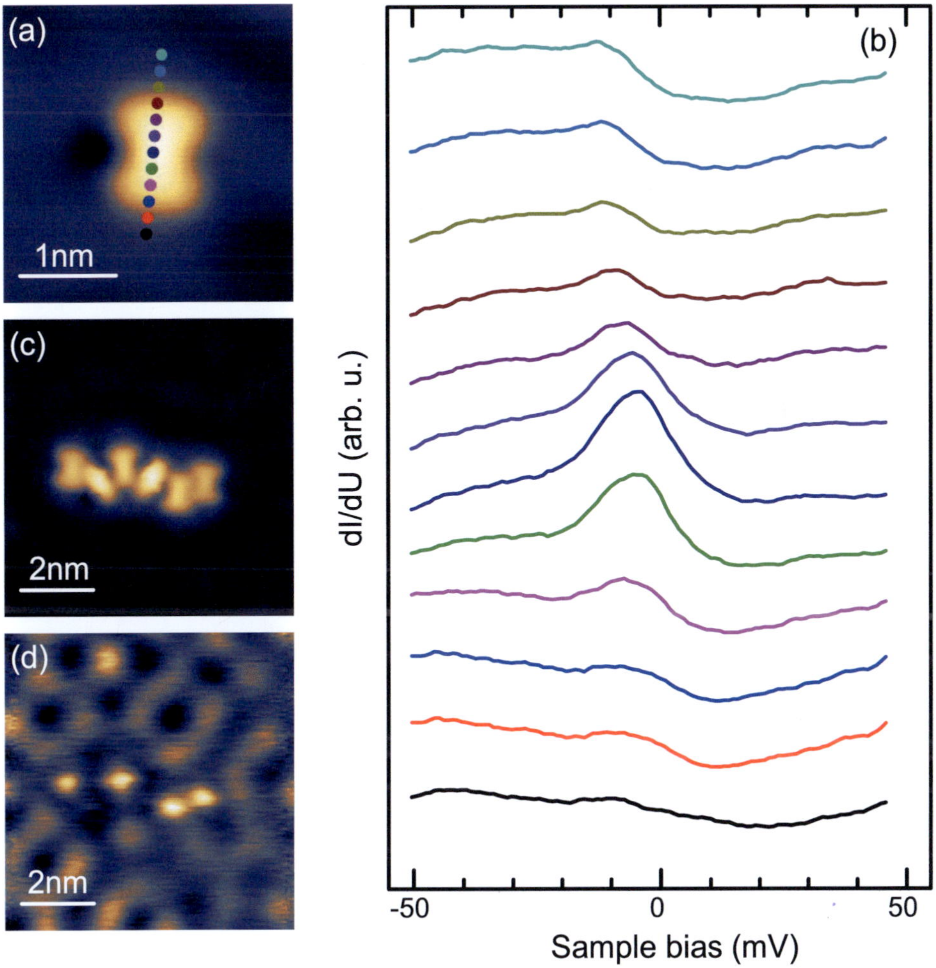

Figure 5.8: STS data were taken as function of positions (a) Topographic image of the molecule and the points where STS were recorded. (b) STS on different points of (a). (c) Topographic image of a molecule chain. (d) dI/dU map of the chain at $U_{bias} = -3$ mV.

5.5 Summary

$Dy(acac)_3$ and $Cr(acac)_3$ molecules were deposited on the Cu(111) surface. Both of them formed self-organized chains on the surface and single molecules can be found among the chains. However $Dy(acac)_3$ and $Cr(acac)_3$ showed different adsorption configurations: single $Dy(acac)_3$ molecules 'lie' on the surface with three acac ligands equally contacting the substrate and shows a threefold symmetry while the $Cr(acac)_3$ molecules 'stand' on the surface with one acac ligand as 'foot'.

The STS studies shows that spin-flip processes arises on the $Dy(acac)_3$ during tunneling and three inelastic excitations can be recognized, two of which should be spin-excitations. A remarkable MAE of $Dy(acac)_3$ on Cu(111) was estimated with a value of 12 meV. In contrast to $Dy(acac)_3$, a clear Kondo peak was resolved in STS of $Cr(acac)_3$ molecule on Cu(111) surface and a Kondo temperature of $T_K = 34 \pm 2$ Kelvin was estimated. As explained in the beginning of this chapter, the origin of this difference is the different values and symmetries of the spin configurations of the Dy(III) and Cr(111) ions, which can influence whether a Kondo resonance can arise in the system [53].

6 Conclusions and outlook

In this work, the magnetic properties, such as the exchange interaction between the magnetic ions, the magnetic anisotropy on the surface, the magnetic excitations as well as the Kondo effect of different magnetic molecules on different substrates were studied using STM.

The entirely homebuilt STM including a ^{3}He Joule-Thomson refrigerator was developed and built to support this work [77]. A base temperature of 650 mK was achieved by using a ^{3}He-^{4}He mixture. An even lower base temperature can be obtained by using pure ^{3}He. Due to the carefully design, the consumption of the cryogenic liquids is very low and a standing time of liquid helium of 280 hours was achieved. This setup has a high mechanical stability and a high energy resolution. The vibration level of the instrument at the STM tip is only several 100 fm and the energy resolution is better than 0.3 meV for STS measurements (dI/dU) at the base temperature. Furthermore a superconducting coil has been constructed, which can apply an out-of-plane magnetic field of 4 T at the STM sample surface. Due to the good performance of the cryostat and the STM, the design of the machine has been licensed to industrial partners and is already on the market.

STM studies have been performed on the three (M-(hfacac)$_2$)$_2$-bpym complexes (M=Ni, Mn, Zn) on the metallic Au(111) or Cu(100) surface as well as Ni_2 on insulating CuN/Cu(100) surface. The original idea of this experiment was to determine the exchange interaction between the two Mn(II)/Ni(II) ions by performing ITS. However, a Kondo resonance arose on both Ni_2 and Mn_2 molecules on the metallic Au(111) and Cu(100) surfaces, which is unexpected in such an integer spin system with two antiferromagnetically coupled magnetic ions. A possible explanation for this phenomenon is that the coupling between the magnetic ions and the conduction electrons in the substrate is stronger than the antiferromagnetic coupling between the two ions, so that the spin of the each single ion is screened and the interaction between the two ions is quenched. The experiment on Ni_2 on the insulating CuN/Cu(100) surface gave an evidence for this interpretation. By reducing the coupling between the magnetic ions and the conduction electrons with the ultrathin insulating CuN layer, the Kondo resonance was suppressed. However, theoretical calculations are still needed to solve this problem and it will be done in the near future by our theoretical partners.

The STS studies on $Dy(acac)_3$ and $Cr(acac)_3$ on Cu(111) showed that the Kondo resonance appeared in the $Cr(acac)_3$/Cu(111) system but not in the $Dy(acac)_3$/Cu(111), just as predicted in Chapter 2. Furthermore a remarkable MAE was revealed in the $Dy(acac)_3$/Cu(111) system. This result proves that by choosing the SMMs and the substrate carefully considering the symmetry of the Hamiltonian of the interaction between SMMs and the substrate, the Kondo effect can be suppressed and the existence of a stable magnetic ground state of SMMs on the metallic surface is possible. This is one step closer to magnetic information storage on SMMs.

Deutsche Zusammenfassung

Wie vom Moore'schen und Kryder'schen Gesetz vorhergesagt [1, 2], ließ sich die Halbleitertechnik in den letzten Jahrzehnten bemerkenswert verbessern und die Flächendichte von magnetischen Speicherelementen schnell erhöhen. Vor kurzem sind Prozessoren auf dem Markt erschienen, die mit Hilfe der 22-nm-Technologie hergestellt wurden [3], sowie Festplatten mit einer Datendichte von 620 GBit pro Quadratzoll [4], was einer Bitgröße von 30 x 30 nm^2 entspricht. Allerdings können diese Entwicklungstrends wegen der physikalischen Grenzen nicht ewig anhalten. In absehbarer Zeit werden Geräte mit Strukturen von der Größenordnung einzelner Atome oder Moleküle benötigt, was die heutige industrielle Technologie jedoch leider nicht erreichen kann. Daher haben magnetische Moleküle, vor allem Einzelmolekülmagnete (engl. single molecule magnets, kurz SMMs), viel Interesse auf sich gezogen. Sie verhalten sich wie einzelne molekulare magnetische Bits [5] und erscheinen vielversprechend für potentielle Anwendungen in der molekularen Spintronik [6], wie beispielsweise Quanten-Computing [7].

Für Anwendungen wie magnetische Datenspeicherung, müssen die magnetischen Moleküle auf ein unterstützendes Substrat aufgedampft werden. Um ein magnetisches Bit auf einer Oberfläche zu realisieren, werden Moleküle benötigt, die eine intrinsische Remanenz und zwei stabile magnetische Grundzuständen besitzen, sowie einer Energiebarriere, der sogenannten magnetischen Anisotropieenergie (MAE), die die beiden Grundzustände trennt. Weiterhin sollte die MAE ausreichend groß sein, sodass die magnetischen Bits genug thermische Stabilität haben. Allerdings können im System aufgrund der Hybridisierung der molekularen Zustände mit den Elektronen des Substrats einige Effekte höherer Ordnung erscheinen, wie zum Beispiel der Kondo-Effekt [8]. Bei diesem Effekt wird der Spin des SMMs durch die Leitungselektronen des Substrats abgeschirmt [9] und der magnetische Grundzustand des SMMs geht verloren. Um Informationen in SMMs zu speichern, sollten deswegen sowohl die Art des SMMs als auch das Substrat sorgfältig ausgewählt werden, sodass der Kondo-Effekt unterdrückt werden kann.

Für die Untersuchung der magnetischen Moleküle auf einem Substrat wird ein Instrument mit hoher lateraler Auflösung und hoher Energieauflösung für elektronische Zustände benötigt. Eine übliche Lösung für diese Anforderung ist ein Rastertunnelmikroskop (STM) [10, 11, 12], das bei tiefen Temperaturen betrieben wird [13, 14, 15, 16, 17, 18, 19, 20]. Für Forschungen auf einer atomar sauberen Oberfläche, muss sich das STM im Ultrahochvakuum (UHV) befinden [12, 21, 22, 23]. Mit Hilfe der Rastertunnelspektroskopie (STS) kann man die lokale Zustandsdichte (LDOS) der untersuchten Objekte erhalten [12]. Darüber hinaus lässt sich die Energiebarriere für inelastische Prozesse, wie zum Beispiel eines Spin-Flips oder der Anregung von Phononen, sowie die MAE mit der inelastischen Tunnelspektroskopie (ITS) bestimmen. Typische, kommerziell verfügbare Tieftemperatur-UHV-STMs verwendet in der Regel einen Helium-Bad-kryostaten und arbeiten bei 4,2 K [24, 25, 26]. Die Energieauflösung der ITS ist bei gegebener Temperatur T auf 5,4 k_BT begrenzt [27] und liegt somit bei nur 1,9 meV bei 4,2 K. Für viele Untersuchung, wie zum Beispiel von magnetischen Anregungen von einzelnen Atomen oder Molekülen [28, 29, 30], ist diese Auflösung nicht ausreichend und es werden niedrigere Temperaturen benötigt, welche durch die Verwendung eines Joule-Thomson-Kryostaten oder eines Mischkryostaten erreicht werden können.

In dieser Arbeit wurde das magnetische Verhalten verschiedener magnetischer Moleküle auf Substraten mit Hilfe der Rastertunnelmikroskopie (STM) untersucht, wobei insbesondere die Austausch-Wechselwirkung zwischen den magnetischen Ionen, die magnetische Anisotropie auf der Oberfläche, magnetische Anregungen, sowie der Kondo-Effekt betrachtet wurden. Der theoretische Hintergrund des Spin-Flip-Prozesses der magnetischen Moleküle im Kristallfeld, der Kondo-Effekt, sowie eine Einführung zu STM und STS sind in Kapitel 2 dargestellt. Das verwendete STM einschließlich des ^{3}He-Joule-Thomson-Kryostaten wurde in Eigenarbeit entwickelt und gebaut [77]. Eine Betriebstemperatur von 650 mK wurde durch die Verwendung eines ^{3}He-^{4}He-Gemisches erreicht, wobei sich noch niedrigere Temperaturen erreichen lassen, wenn die Mischung durch reines ^{3}He ersetzt wird. Aufgrund des durchdachten Designs ist der Verbrauch der kryogenen Flüssigkeiten sehr gering und eine Standzeit von flüssigem Helium von 280 Stunden wurde erreicht. Die Apparatur hat eine hohe mechanische Stabilität und eine hohe Energieauflösung. Die Amplitude der Schwingung an der STM-Spitze beträgt nur einige 100 fm und die Energieauflösung ist besser als 0,3 meV für STS-Messungen (dI/dU) bei 650 mK. Darüber hinaus wurde eine supraleitende Spule konstruiert, um ein senkrecht zur Probenoberfläche orientiertes Magnetfeld von 4 T zu erzeugen. Eine Einführung der gesamten Apparatur wird in Kapitel 3 gegeben.

Magnetische Moleküle enthalten oft mehr als ein magnetisches Zentrum, welche ferromagnetisch oder antiferromagnetisch über die Austauschwechselwirkung gekoppelt sind [31]. Im einfachsten Fall mit antiferromagnetisch gekoppelten Zentren enthalten die Moleküle zwei magnetische Ionen. In Kapitel 4 wird eine Serie von zweikernigen Metall-organischen Molekülen der Form (M(hexafluoracetylacetonat)$_2$)$_2$-Bipyrimidin (kurz M_2), wobei M = Nickel, Mangan, Zink, untersucht, um die antiferromagnetische Kopplung zwischen den beiden übergangsmetallionen zu messen. Allerdings zeigen sowohl Ni_2 als auch Mn_2 auf den metallischen Oberflächen von Au(111) beziehungsweise Cu(100) eine Kondo-Resonanz, was in einem solchen ganzzahligen Spin-System mit zwei antiferromagnetisch gekoppelten magnetischen Ionen unerwartet ist. Eine mögliche Erklärung für dieses Phänomen ist, dass die Kopplung zwischen den magnetischen Ionen und den Leitungselektronen in dem Substrat stärker als die antiferromagnetische Kopplung zwischen den beiden Ionen ist, sodass jeweils die magnetischen Momente der einzelnen Ionen abgeschirmt werden und die Wechselwirkung zwischen den beiden Ionen unterdrückt wird. Das Experiment an Ni_2 auf einer isolierenden CuN/Cu(100) Oberfläche ergab einen Beweis für diese Vermutung. Durch Reduzieren der Kopplung zwischen den magnetischen Ionen und den Leitungselektronen durch die ultradünne isolierende CuN-Schicht wird die Kondo-Resonanz unterdrückt. Jedoch fehlt derzeit noch eine theoretische Berechnung, um diese Erklärung zu stützen.

Die Metallacetylacetonate sind eine Reihe von Koordinationskomplexen, die intensiv untersucht wurden [32, 33, 34, 35, 36, 37, 38, 39, 40]. Dabei zeigte Dysprosium-Acetylacetonat (Dy(acac)$_3$) einen Relaxationsseffekt entsprechend einer Spin-Gitter-Relaxationszeit von etwa einer Millisekunde, die typisch für SMMs ist. Allerdings sind die Untersuchung über das Wachstum und die Eigenschaften der Metallacetylacetonaten auf Oberflächen noch selten. In Kapitel 5 werden STM-Untersuchungen an Dy(acac)$_3$ und Chrom-Acetylacetonat (Cr(acac)$_3$) vorgestellt. Die STS-Messungen an Dy(acac)$_3$ und Cr(acac)$_3$ auf Cu(111) zeigten, dass die Kondo-Resonanz im Cr(acac)$_3$/Cu(111)-System erschient, aber nicht in Dy(acac)$_3$/Cu(111), genau wie in Kapitel 2 vorhergesagt. Außerdem wurde eine bemerkenswerte MAE im Dy(acac)$_3$/Cu(111)-System gefunden. Dieses Ergebnis beweist, dass durch die vorsichtige Wahl der SMMs und des Substrates mit Berücksichtigung der Symmetrie des Hamilton-Operators der Wechselwirkung zwischen SMM und dem Substrat der Kondo-Effekt unterdrückt werden kann und die Existenz stabiler magnetischer Grundzustände von SMMs auf metallischen Oberflächen möglich ist. Dies ist ein Schritt näher an das Ziel, magnetische Speicherung von Informationen mit Hilfe von SMMs zu ermöglichen.

Bibliography

[1] G.E. Moore. Cramming more components onto integrated circuits. *Proceedings of the IEEE*, 86(1):82–85, January 1998.

[2] C. Walter. Kryder's law. *Scientific American*, 32:293, 2005.

[3] www.intel.com. 2012.

[4] www.seagate.com. 2012.

[5] Matteo Mannini, Francesco Pineider, Philippe Sainctavit, Chiara Danieli, Edwige Otero, Corrado Sciancalepore, Anna Maria Talarico, Marie-Anne Arrio, Andrea Cornia, Dante Gatteschi, and Roberta Sessoli. Magnetic memory of a single-molecule quantum magnet wired to a gold surface. *Nature Materials*, 8:194–197, 2009.

[6] Lapo Bogani and Wolfgang Wernsdorfer. Molecular spintronics using single-molecule magnets. *Nature Materials*, 7:179–186, 2008.

[7] M. N. Leuenberger and D. Loss. Quantum computing in molecular magnets. *Nature*, 410:789–793, 2001.

[8] Wenjie Liang, Matthew P. Shores, Marc Bockrath, Jeffrey R. Long, and Hongkun Park. Kondo resonance in a single-molecule transistor. *Nature*, 417:725–729, 2002.

[9] Kenneth G. Wilson. The renormalization group: Critical phenomena and the kondo problem. *Rev. Mod. Phys.*, 47:773–840, Oct 1975.

[10] G. Binnig, H. Rohrer, Ch. Gerber, and E. Weibel. Tunneling through a controllable vacuum gap. *Appl. Phys. Lett.*, 40:178–180, 1982.

[11] G. Binnig, H. Rohrer, Ch. Gerber, and E. Weibel. Surface studies by scanning tunneling microscopy. *Phys. Rev. Lett.*, 49(1):57–61, Jul 1982.

[12] C.J. Chen. *Introduction to Scanning Tunneling Microscopy*. Oxford University Press, 1993.

[13] A. L. de Lozanne, S. A. Elrod, and C. F. Quate. Spatial variations in the superconductivity of Nb_3Sn measured by low-temperature tunneling microscopy. *Phys. Rev. Lett.*, 54(22):2433–2436, Jun 1985.

[14] A. P. Fein, J. R. Kirtley, and R. M. Feenstra. Scanning tunneling microscope for low temperature, high magnetic field, and spatially resolved spectroscopy. *Rev. Sci. Instrum.*, 58(10):1806–1810, 1987.

[15] D. M. Eigler and E. K. Schweizer. Positioning single atoms with a scanning tunnelling microscope. *Nature*, 344:524 – 526, 1990.

[16] P. Davidsson, H. Olin, M. Persson, and S. Pehrson. Design and operation of a low-temperature scanning tunneling microscope suitable for operation below 1 K. *Ultramicroscopy*, 42-44:1470 – 1475, 1992.

[17] S. H. Tessmer, D. J. Van Harlingen, and J. W. Lyding. Integrated cryogenic scanning tunneling microscopy and sample preparation system. *Rev. Sci. Instrum.*, 65:2855, 1994.

[18] J. W. G. Wildöer, A. J. A. van Roy, H. van Kempen, and C. J. P. M. Harmans. Low-temperature scanning tunneling microscope for use on artificially fabricated nanostructures. *Rev. Sci. Instrum.*, 65:2849, 1994.

[19] S. H. Pan, E. W. Hudson, and J. C. Davis. ^{3}He refrigerator based very low temperature scanning tunneling microscope. *Rev. Sci. Instrum.*, 70(2):1459–1463, 1999.

[20] Young Jae Song, Alexander F. Otte, Vladimir Shvarts, Zuyu Zhao, Young Kuk, Steven R. Blankenship, Alan Band, Frank M. Hess, and Joseph A. Stroscio. A 10 mK scanning probe microscopy facility. *Rev. Sci. Instrum.*, 81:121101, 2010.

[21] Joseph A. Stroscio and D. M. Eigler. Atomic and Molecular Manipulation with the Scanning Tunneling Microscope. *Science*, 254(5036):1319–1326, 1991.

[22] A. J. Heinrich, C. P. Lutz, J. A. Gupta, and D. M. Eigler. Molecule cascades. *Science*, 298(5597):1381–1387, Nov 2002.

[23] L. Gerhard, T. K. Yamada, T. Balashov, A. F. Takács, R. J. H. Wesselink, M. Däne, M. Fechner, S. Ostanin, A. Ernst, I. Mertig, and W. Wulfhekel. Magnetoelectric coupling at metal surfaces. *Nature Nanotechnology*, 5:792–797, 2010.

[24] *RHK Technology, USA.*

[25] *Omicron NanoTechnology GmbH, Germany.*

[26] *Unisoku Co., Ltd, Japan.*

[27] J. Klein, A. Léger, M. Belin, D. Défourneau, and M. J. L. Sangster. Inelastic-electron-tunneling spectroscopy of metal-insulator-metal junctions. *Phys. Rev. B*, 7(6):2336–2348, Mar 1973.

[28] T. Balashov, A. F. Takács, W. Wulfhekel, and J. Kirschner. Magnon excitation with spin-polarized scanning tunneling microscopy. *Phys. Rev. Lett.*, 97:287201, 2006.

[29] T. Balashov, T. Schuh, A. F. Takacs, A. Ernst, S. Ostanin, J. Henk, I. Mertig, P. Bruno, T. Miyamachi, S. Suga, and W. Wulfhekel. Magnetic anisotropy and magnetization dynamics of individual atoms and clusters of Fe and Co on Pt(111). *Phys. Rev. Lett.*, 102:257203, 2009.

[30] T. Schuh, T. Balashov, T. Miyamachi, A. F. Takacs, S. Suga, and W. Wulfhekel. Lifetimes of magnetic excitations in Fe and Co atoms and clusters on Pt(111). *J. App. Phys.*, 107:9E156, 2010.

[31] D. Gatteschi, R. Sessoli, and J. Villain. *Molecular nanomagnets*. Oxford Univ Press, 2006.

[32] Ingmar Grenthe and W. Conard Fernelius. Stability relationships among the rare earth acetylacetonates. *J. Am. Chem. Soc.*, 82:6258–6260, 1960.

[33] G.W. Pope, J.F. Steinbach, and W.F. Wagner. Characteristics of the solvates of the rare-earth acetylacetonates. *Journal of Inorganic and Nuclear Chemistry*, 20:304–313, 1961.

[34] J. J. Fritz, P. E. Field, and Ingmar Grenthe. The low temperature magnetic properties and magnetic energy levels of some rare earth chelates of acetylacetone and ethylenediaminetetraacetic acid. *J. Phys. Chem.*, 65:2070–2074, 1961.

[35] W.B. Brown, G.W. Pope, J.F. Steinbach, and W.F. Wagner. Countercurrent extraction of the rare-earth acetylacetonates. *Journal of Inorganic and Nuclear Chemistry*, 25:429–439, 1968.

[36] D. R. Eaton. The nuclear magnetic resonance of some paramagnetic transition metal acetylacetonates. *J. Am. Chem. Soc.*, 87:3097–3102, 1965.

[37] Mary Frances Richardson, William F. Wagner, and Donald E. Sands. Anhydrous and hydrated rare earth acetylacetonates and their infrared spectra. *Inorg. Chem.*, 7:2495–2500, 1968.

[38] Ivan B. Liss and William G. Bos. Rare earth acetylacetonates: New preparative methods and new complexes. *Journal of Inorganic and Nuclear Chemistry*, 39:443–447, 1977.

[39] L. I. Martynenko, I. A. Murav'eva, N. K. Khalmurzaev, and V. I. Spitsyn. Synthesis and IR spectroscopic study of rare-earth tris(acetylacetonates). *Russian Chemical Bulletin*, 33:449–454, 1984.

[40] M. Aslam Siddiqi, Rehan A. Siddiqui, and Burak Atakan. Thermal stability, sublimation pressures, and diffusion coefficients of anthracene, pyrene, and some metal β-diketonates. *J. Chem. Eng. Data*, 54:2795–2802, 2009.

[41] Stefan Schmaus. *Spintronics with individual metal-organic molecules*. PhD thesis, Karlsruhe Institute of Technology, 2011.

[42] G. Binnig and H. Rohrer. Scanning tunneling microscopy. *Helv. Phys. Acta*, 55:726, 1982.

[43] J. Bardeen. Tunnelling from a many-particle point of view. *Phys. Rev. Lett.*, 6:57–59, Jan 1961.

[44] J. Tersoff and D. R. Hamann. Theory and application for the scanning tunneling microscope. *Phys. Rev. Lett.*, 50:1998–2001, Jun 1983.

[45] J. Tersoff and D. R. Hamann. Theory of the scanning tunneling microscope. *Phys. Rev. B*, 31:805–813, Jan 1985.

[46] Vladimir A. Ukraintsev. Data evaluation technique for electron-tunneling spectroscopy. *Phys. Rev. B*, 53:11176–11185, Apr 1996.

[47] S. H. Pan. *Piezo-electric Motor, International Patent Publication No. WO 93/19494, International Bureau, World Intellectual Property Organization*, 30 September 1993.

[48] B. C. Stipe, M. A. Rezaei, and W. Ho. Single-molecule vibrational spectroscopy and microscopy. *Science*, 280:1732–1735, 1998.

[49] W. Olejniczak, Z. Klusek, and M. Bieniecki. Investigations of the fine structure of I(V) characteristics for highly oriented pyrolitic graphite surface by means of STM/STS at room temperature. *Applied Physics A*, 66:191–196, 1998.

[50] L. Vitali, M. A. Schneider, K. Kern, L. Wirtz, and A.Rubio. Phonon and plasmon excitation in inelastic electron tunneling spectroscopy of graphite. *Physical Review B*, 69:121414–121417, 2004.

[51] H. Gawronski, M. Mehlhorn, and K. Morgenstern. Single-molecule vibrational spectroscopy and microscopy. *Science*, 319:930–933, 2008.

[52] Tobias Schuh, Timofey Balashov, Toshio Miyamachi, Shih-Yu Wu, Chein-Cheng Kuo, Arthur Ernst, Jürgen Henk, and Wulf Wulfhekel. Magnetic anisotropy and magnetic excitations in supported atoms. *Phys. Rev. B*, 84:104401, Sep 2011.

[53] Tobias Schuh. *Spin-Dynamik in Atomen und Clustern auf Oberflächen*. PhD thesis, Karlsruhe Institute of Technology, 2011.

[54] D. Gatteschi and R. Sessoli. Quantum tunneling of magnetization and related phenomena in molecular materials. *Angew. Chem. Int. Ed.*, 42:268–297, 2003.

[55] W. J. de Haas and G. J. van den Berg. The electrical resistance of gold and silver at low temperatures. *Physica III*, 6:440, 1936.

[56] Jun Kondo. Resistance minimum in dilute magnetic alloys. *Progress of Theoretical Physics*, 32:37, 1964.

[57] P. W. Anderson. Localized magnetic states in metals. *Phys. Rev.*, 124:41–53, Oct 1961.

[58] L. Kouwenhoven and L. Glazman. Revival of the Kondo effect. *Physics World*, 14(1):33–38, 2001.

[59] D. Goldhaber-Gordon, Hadas Shtrikman, D. Mahalu, David Abusch-Magder, U. Meirav, and M. A. Kastner. Kondo effect in a single-electron transistor. *Nature*, 391(6663):156–159, January 1998.

[60] S. M. Cronenwett, T. H. Oosterkamp, and L. P. Kouwenhoven. A tunable kondo effect in quantum dots. *Science*, 281:540–544, 1998.

[61] V. Madhavan, W. Chen, T. Jamneala, M. F. Crommie, and N. Wingreen. Tunneling into a single magnetic atom: spectroscopic evidence of the Kondo resonance. *Science*, 280:567–569, 1998.

[62] Jiutao Li, Wolf-Dieter Schneider, Richard Berndt, and Bernard Delley. Kondo scattering observed at a single magnetic impurity. *Phys. Rev. Lett.*, 80:2893–2896, Mar 1998.

[63] H. C. Manoharan, C. P. Lutz, and D. M. Eigler. Quantum mirages formed by coherent projection of electronic structure. *Nature*, 403(6769):512–515, 2000.

[64] P. Wahl, L. Diekhöner, G. Wittich, L. Vitali, M. A. Schneider, and K. Kern. Kondo effect of molecular complexes at surfaces: Ligand control of the local spin coupling. *Phys. Rev. Lett.*, 95:166601, Oct 2005.

[65] P. Wahl, P. Simon, L. Diekhöner, V. S. Stepanyuk, P. Bruno, M. A. Schneider, and K. Kern. Exchange interaction between single magnetic adatoms. *Phys. Rev. Lett.*, 98:056601, Jan 2007.

[66] Alexander F. Otte, Markus Ternes, Kirsten von Bergmann, Sebastian Loth, Harald Brune, Christopher P. Lutz, Cyrus F. Hirjibehedin, and Andreas J. Heinrich. The role of magnetic anisotropy in the Kondo effect. *Nat. Phys.*, 4(11):847–850, November 2008.

[67] Aidi Zhao, Qunxiang Li, Lan Chen, Hongjun Xiang, Weihua Wang, Shuan Pan, Bing Wang, Xudong Xiao, Jinlong Yang, J. G. Hou, and Qingshi Zhu. Controlling the Kondo Effect of an Adsorbed Magnetic Ion Through Its Chemical Bonding. *Science*, 309:1542–1544, 2005.

[68] Keiichi Katoh, Yusuke Yoshida, Masahiro Yamashita, Hitoshi Miyasaka, Brian K Breedlove, Takashi Kajiwara, Shinya Takaishi, Naoto Ishikawa, Hironari Isshiki, Yan Feng Zhang, Tadahiro Komeda, Masakazu Yamagishi, and Jun Takeya. Direct observation of Lanthanide(III)-Phthalocyanine molecules on Au(111) by using scanning tunneling microscopy and scanning tunneling spectroscopy and thin-film field-effect transistor properties of Tb(III) and Dy(III)-Phthalocyanine Molecules. *J. Am. Chem. Soc.*, 131:9967–9976, 2009.

[69] U. Fano. Effects of configuration interaction on intensities and phase shifts. *Phys. Rev.*, 124:1866–1878, Dec 1961.

[70] O. Újsághy, J. Kroha, L. Szunyogh, and A. Zawadowski. Theory of the Fano resonance in the STM tunneling density of states due to a single Kondo impurity. *Phys. Rev. Lett.*, 85:2557–2560, Sep 2000.

[71] J. Merino and O. Gunnarsson. Simple model for scanning tunneling spectroscopy of noble metal surfaces with adsorbed kondo impurities. *Phys. Rev. B*, 69:115404, Mar 2004.

[72] M. Plihal and J. W. Gadzuk. Nonequilibrium theory of scanning tunneling spectroscopy via adsorbate resonances: Nonmagnetic and Kondo impurities. *Phys. Rev. B*, 63:085404, Feb 2001.

[73] K. Nagaoka, T. Jamneala, M. Grobis, and M. F. Crommie. Temperature dependence of a single Kondo impurity. *Phys. Rev. Lett.*, 88:077205, Feb 2002.

[74] D. Goldhaber-Gordon, J. Göres, M. A. Kastner, Hadas Shtrikman, D. Mahalu, and U. Meirav. From the Kondo regime to the mixed-valence regime in a single-electron transistor. *Phys. Rev. Lett.*, 81:5225–5228, Dec 1998.

[75] J. J. Parks, A. R. Champagne, T. A. Costi, W. W. Shum, A. N. Pasupathy, E. Neuscamman, S. Flores-Torres, P. S. Cornaglia, A. A. Aligia, C. A. Balseiro, G. K.-L. Chan, H. D. Abruna, , and D. C. Ralph. Mechanical control of spin states in spin-1 molecules and the underscreened Kondo effect. *Science*, 328:1370–1373, 2010.

[76] U. Schlickum, W. Wulfhekel, and J. Kirschner. A spin-polarized scanning tunneling microscope for imaging the in-plane magnetization. *Appl. Phys. Lett.*, 83:2016, 2003.

[77] L. Zhang, T. Miyamachi, T. Tomani, R. Dehm, and W. Wulfhekel. A compact sub-Kelvin ultrahigh vacuum scanning tunneling microscope with high energy resolution and high stability. *Rev. Sci. Instrum.*, 82:103702, 2011.

[78] Fank Pobell. *Matter and Methods at Low Temperatures*. Springer-Verlag, 1996.

[79] Jack W. Ekin. *Experimental Techniques for Low-Temperature Measurements*. Oxford University Press, 2006.

[80] *Lake Shore Cryotronics, USA*.

[81] F G Brickwedde and United States. National Bureau of Standards. *The "1958 He^4 Scale of Temperatures"*. Washington, 1960.

[82] C. F. Barenghi, R. J. Donnelly, and R. N. Hills. Cubic Spline Fit to the 1958 ^{4}He Scale of Temperatures. *Journal of Low Temperature Physics*, 51:319–327, 1982.

[83] L. D. Landau and E. M. Lifshitz. *Statistical physics*. Pergamon Press, 1970.

[84] J.Kraus, E. Uhlig, and W. Wiedemann. Enthalpy-pressure($H - p$) diagram of He^3 in the range 1.0 K $\leq T \leq$ 4.17 K and $0 \leq p \leq 6.5$ atm and inversion curve for $T \leq 4.17$ K. *Cryogenics*, 14(1):29–35, 1974.

[85] John E. Rives and Horst Meyer. Density of liquid Helium-3 between 0.045 K und 1.3 K. *Phys. Rev. Letters*, 7(6):217–219, 1961.

[86] D. S. Betts, D. W. Osborne, B. Welber, and J. Wilks. The viscosity of liquid helium 3. *Philosophical Magazine*, 90(8):977–987, 1963.

[87] William E. Keller. Calculation of the viscosity of gaseous He^3 and He^4 at low temperatures. *Phys. Rev.*, 105:41–45, Jan 1957.

[88] William F. Hughes and John A. Brighton. *Schaum's outline of theory and problems of fluid dynamics*. McGraw Hill, 1999.

[89] Alfred H. Grunberg, L.and Nissan. Mixture law for viscosity. *Nature*, 164:799–800, 1949.

[90] *Nanonis Control System, SPECS Surface Nano Analysis GmbH, Germany, www.specs.com.*

[91] *Variable Gain Low Noise Current Amplifier DLPCA-200, FEMTO Messtechnik GmbH, Germany.*

[92] J. V. Barth, H. Brune, G. Ertl, and R. J. Behm. Scanning tunneling microscopy observations on the reconstructed Au(111) surface: Atomic structure, long-range superstructure, rotational domains, and surface defects. *Phys. Rev. B*, 42(15):9307–9318, Nov 1990.

[93] A. A. Abrikosov and Zh. Eksp. On the magnetic properties of superconductors of the second group. *Teor. Fiz.*, 32:1442, 1957.

[94] Michael Tinkham. *Introduction to Superconductivity: Second Edition*. McGraw-Hill, 1996.

[95] George Christou, Dante Gatteschi, David N. Hendrickson, and Roberta Sessoli. Single-molecule magnets. *Mater. Res. Soc. Bull.*, 25:66–71, 2000.

[96] Massimiliano Cavallini, Massimo Facchini, Cristiano Albonetti, and Fabio Biscarini. Single molecule magnets: from thin films to nano-patterns. *Physical Chemistry Chemical Physics*, 10:784–793, 2008.

[97] Montserrat Barquín, María J. González Garmendia, and Víctor Bellido. Homo- and heterodinuclear 2,2'-bipyrimidine-bridged complexes. *Transition Metal Chemistry*, 24:546–552, 1999.

[98] Frank Schramm and Mario Ruben. *personal communication*, 2009.

[99] G. Brewer and E. Sinn. 2,2'-bipyrimidine-bridged homobinuclear complexes. *Inorganic Chemistry*, 24:4580–4584, 1985.

[100] Avraham Schiller and Selman Hershfield. Theory of scanning tunneling spectroscopy of a magnetic adatom on a metallic surface. *Phys. Rev. B*, 61:9036–9046, Apr 2000.

[101] M. N. Kiselev. Dynamical symmetries and quantum transport through nanostructures. *physica status solidi (c)*, 4:3362, 2007.

[102] Aitor Mugarza, Cornelius Krull, Roberto Robles, Sebastian Stepanow, Gustavo Ceballos, and Pietro Gambardella. Spin coupling and relaxation inside molecule-metal contacts. *Nat Commun.*, 24:490, 2011.

[103] Gary G. Tibbetts, James M. Burkstrand, and J. Charles Tracy. Electronic properties of adsorbed layers of nitrogen, oxygen, and sulfur on copper (100). *Phys. Rev. B*, 15:3652–3660, Apr 1977.

[104] C. D. Ruggiero, T. Choi, and J. A. Gupta. Tunneling spectroscopy of ultrathin insulating films: CuN on Cu(100). *Appl. Phys. Lett.*, 91:253106, 2007.

[105] Z. Szafran, R.M. Pike, and M.M. Singh. *Microscale Inorganic Chemistry: A Comprehensive Laboratory Experience*. John Wiley & Sons, 1991.

[106] Timofey Balashov. *Inelastic scanning tunneling spectroscopy: magnetic excitations on the nanoscale*. PhD thesis, Karlsruhe Institute of Technology, 2009.

Acknowledgements

First of all, I would like to thank Prof. Dr. Wulf Wulfhekel forthe opportunity for the promotion, the conscientious supervising, enduring patience and all kinds of support he has given to me in the last five years. His deep insights into physics, like a torch in the darkness, always guides me in the right direction.

I would also like to thank Prof. Dr. Georg Weiß for agreeing to be the second referee, and for investing time and effort into reading and commenting on this work.

I am grateful to Prof. Dr. Mario Ruben and Dr. Frank Schramm for the synthesis and essential information about the molecules.

I am also grateful to Prof. Dr. Hilbert von Löhneysen for the generous ^3He.

I feel also deeply grateful to Prof. Dr. Saw-Wai Hla who allowed me to use the only working STM in his lab at that time for whole three months. I also want to thank all his group members who gave me a warm welcome and lots of support during my stay in US.

I want to thank the Karlsruhe House of Young Scientists (KHYS) for the research travel scholarship which supported me to accomplish the three-months-trip to US as a short term scholar.

Special thanks to Dr. Toshio Miyamachi for teaching me lots of physics, for constant help in the experiments, all his painstaking night shifts and hard work during holidays. He gave me also nice and professional suggestions after reading my manuscript.

I want to thank Michael Schackert for investing a lot of time to erase as many mistakes as possible in this thesis.

I also want to thank Tobias Schuh for teaching me the stuff about the MAE, Stevens operators and skiing.

I am grateful to the whole group including the former members for the enjoyable working atmosphere.

Special thanks the staff of the mechanical workshop, electronic workshop for their cooperative, excellent work. They manufactured the whole setup.

I would like to thank Dr Christoph Sürgers who always lent me all kinds of stuff: spot welding machine, Helium leak detector, Hall sensor, wobble stick, platinum electrode, ... even copper gaskets.

I would also like to thank Torben Peichl for the calibration of the Cernox thermometer.

I want to thank Ms Hugle for cutting the crystal and ceramics as well as preparing the chemical experiments.

I would like to thank all my friends for the whole time in Karlsruhe.

Thanks to my parents for all the support.

Finally, thanks to Xiaoxin for everything.

Experimental Condensed Matter Physics
(ISSN 2191-9925)

Herausgeber
Physikalisches Institut

Prof. Dr. David Hunger
Prof. Dr. Alexey Ustinov
Prof. Dr. Georg Weiß
Prof. Dr. Wolfgang Wernsdorfer
Prof. Dr. Wulf Wulfhekel

Band 1 Alexey Feofanov
Experiments on flux qubits with pi-shifters. 2011
ISBN 978-3-86644-644-1

Band 2 Stefan Schmaus
Spintronics with individual metal-organic molecules. 2011
ISBN 978-3-86644-649-6

Band 3 Marc Müller
Elektrischer Leitwert von magnetostriktiven Dy-Nanokontakten. 2011
ISBN 978-3-86644-726-4

Band 4 Torben Peichl
Einfluss mechanischer Deformation auf atomare Tunnelsysteme – untersucht mit Josephson Phasen-Qubits. 2012
ISBN 978-3-86644-837-7

Band 5 Dominik Stöffler
Herstellung dünner metallischer Brücken durch Elektromigration und Charakterisierung mit Rastersondentechniken. 2012
ISBN 978-3-86644-843-8

Band 6 Tihomir Tomanic
Untersuchung des elektronischen Oberflächenzustands von Ag-Inseln auf supraleitendem Niob (110). 2012
ISBN 978-3-86644-898-8

Band 7 Lukas Gerhard
Magnetoelectric coupling at metal surfaces. 2013
ISBN 978-3-7315-0063-6

Band 8 Nicht erschienen.

Die Bände sind unter www.ksp.kit.edu als PDF frei verfügbar oder als Druckausgabe bestellbar.

Band 9 Jochen Zimmer
Cooper pair transport in arrays of Josephson junctions. 2014
ISBN 978-3-7315-0130-5

Band 10 Oliver Berg
Elektrischer Transport durch Nanokontakte von Selten-Erd-Metallen. 2014
ISBN 978-3-7315-0209-8

Band 11 Grigorij Jur'evic Grabovskij
Investigation of coherent microscopic defects inside the tunneling barrier of a Josephson junction. 2014
ISBN 978-3-7315-0210-4

Band 12 Cornelius Thiele
STM Characterization of Phenylene-Ethynylene Oligomers on Au(111) and their Integration into Carbon Nanotube Nanogaps. 2014
ISBN 978-3-7315-0235-7

Band 13 Michael Peter Schackert
Scanning Tunneling Spectroscopy on Electron-Boson Interactions in Superconductors. 2014
ISBN 978-3-7315-0238-8

Band 14 Susanne Butz
One-Dimensional Tunable Josephson Metamaterials. 2014
ISBN 978-3-7315-0271-5

Band 15 Philipp Jung
Nonlinear Effects in Superconducting Quantum Interference Meta-Atoms. 2014
ISBN 978-3-7315-0294-4

Band 16 Sebastian Probst
Hybrid quantum system based on rare earth doped crystals. 2016
ISBN 978-3-7315-0345-3

Band 17 Wolfram Kittler
Magnetische Anisotropie und Quantenphasenübergang in $CeTi_{1-x}V_xGe_3$. 2015
ISBN 978-3-7315-0363-7

Band 18 Moritz Peter
Towards magnetic resonance in scanning tunneling microscopy using heterodyne detection. 2015
ISBN 978-3-7315-0410-8

Band 19 Junji Tobias Märkl
Investigation of Magnetic Adatoms with Scanning Tunneling Techniques. 2015
ISBN 978-3-7315-0435-1

Band 20 Jochen Braumüller
Quantum simulation experiments with superconducting circuits. 2018
ISBN 978-3-7315-0780-2

Band 21 Jinjie Chen
Local Investigation of Single Magnetic Molecules with Scanning Tunneling Microscopy. 2018
ISBN 978-3-7315-0819-9

Band 22 Arnold Seiler
Einfluss der Leitungselektronen auf die Dynamik atomarer Tunnelsysteme in ungeordneten Festkörpern: Relaxationsprozesse in metallischen Gläsern und ungeordneten dünnen Aluminiumoxid-Schichten. 2019
ISBN 978-3-7315-0870-0

Band 23 Jasmin Maria Jandke
Elastic and Inelastic Scanning Tunneling Spectroscopy on Iron-Based Superconductors. 2019
ISBN 978-3-7315-0747-5

Band 24 Kevin Edelmann
Electroluminescence from Plasmonic Excitations in a Scanning Tunnelling Microscope. 2019
ISBN 978-3-7315-0923-3

Band 25 Lei Zhang
Sub-Kelvin scanning tunneling microscopy on magnetic molecules. 2019
ISBN 978-3-86644-950-3

Die Bände sind unter www.ksp.kit.edu als PDF frei verfügbar oder als Druckausgabe bestellbar.